Vortex Dynamics - Theoretical, Experimental and Numerical Approaches

Edited by Naoto Ohmura

Published in London, United Kingdom

Vortex Dynamics - Theoretical, Experimental and Numerical Approaches
http://dx.doi.org/10.5772/intechopen.1003392
Edited by Naoto Ohmura

Contributors
Andrew Kaminsky, Andrew Otto, Andrey Aksenov, Emily Friederick, Hayato Masuda, Isaac Ozolins, Ivan Ngian, Jiawen Li, Kelly S. Verrall, Naoto Ohmura, Pearl Scallon, Reagen McCormick, Sergey Timushev, Steven Verrall, Vladimir I. Semenov, Vladimir Shalaev

Notice
Statements and opinions expressed in the chapters are these of the individual contributors and not necessarily those of the editors or publisher. No responsibility is accepted for the accuracy of information contained in the published chapters. The publisher assumes no responsibility for any damage or injury to persons or property arising out of the use of any materials, instructions, methods or ideas contained in the book.

First published in London, United Kingdom, 2025 by IntechOpen
IntechOpen is the global imprint of INTECHOPEN LIMITED, registered in England and Wales, registration number: 11086078, 167-169 Great Portland Street, London, W1W 5PF, United Kingdom

For EU product safety concerns: IN TECH d.o.o., Prolaz Marije Krucifikse Kozulić 3, 51000 Rijeka, Croatia, info@intechopen.com or visit our website at intechopen.com.

British Library Cataloguing-in-Publication Data
A catalogue record for this book is available from the British Library

Vortex Dynamics - Theoretical, Experimental and Numerical Approaches
Edited by Naoto Ohmura
p. cm.
Print ISBN 978-1-83769-814-1
Online ISBN 978-1-83769-813-4
eBook (PDF) ISBN 978-1-83769-815-8

If disposing of this product, please recycle the paper responsibly.

Meet the editor

Naoto Ohmura earned a Ph.D. in Chemical Engineering from Kobe University in 1997 under the supervision of Professor Kunio Kataoka, who is now a Professor Emeritus at Kobe University in Japan. He began his academic career as an Assistant Professor in the Department of Chemical Science and Engineering at Kobe University in 1990. He was subsequently promoted to Associate Professor in 1999 and became a Full Professor in 2004. From 2019 to 2021, he served as the Dean of the Graduate School of Engineering at Kobe University, followed by his tenure as Executive Vice President in charge of education and internationalization from 2021 to 2024. His research interests span across transport phenomena, nonlinear phenomena, mixing and process intensification.

Contents

Preface

The structure and dynamic properties of vortices not only present intriguing mathematical challenges in fluid motion but also play a crucial role in industrial applications, including mixing, heat and mass transfer enhancement. The study of vortices spans multiple scales, from the microscopic behaviors observed in quantum fluids to the macroscopic phenomena governing aerodynamics, atmospheric dynamics, and industrial processes. Organized vortex structures exhibit highly attractive properties, such as solid accumulation, mixing and reaction enhancement, particle classification, and mass transport. To fully harness these properties, it is essential to manipulate vortex structures in a controlled manner. This leads to a fundamental academic question: *Can we control the generation, dissipation, size, shape, and dynamics of vortices?* Addressing this question requires a combination of mathematical, experimental, and numerical approaches to gain deeper insights into vortex motion.

This book, *Vortex Dynamics – Theoretical, Experimental and Numerical Approaches*, brings together diverse perspectives on vortex behavior by addressing both classical and emerging topics in the field. The five chapters in this volume explore a wide range of vortex-related phenomena, each contributing to the broader understanding of fluid dynamics uniquely.

The first chapter proposes a framework for geometrically explaining the proton's mass and charge distribution using the novel concept of a nested surface vortex. By incorporating the Unruh and Zitterbewegung effects, it suggests that the mass of a ground-state proton may be distributed on a spherical surface, offering new insights into the integration of quantum mechanics and relativity.

The second chapter presents a new mathematical approach to understanding and controlling the formation, persistence, and dissipation of vortices through the regularity of solutions to the Navier-Stokes equations and an optimal control framework. It examines the relationship between the smoothness properties of weak solutions beyond a known regularity point and the optimality conditions in the Bellman principle.

The third chapter presents a new approach to describing asymmetric vortex states in separated flows over slender bodies. Using catastrophe theory, it establishes a criterion for the onset of asymmetry in conical bodies and examines key transition properties through local flow analysis. Numerical simulations evaluate the effectiveness of flow control methods such as plasma discharge and surface heating, with experimental validation confirming the results.

The fourth chapter introduces an acoustic-vortex decomposition method to analyze pressure pulsations and noise emissions from blade machines. Separating the velocity field into the vortex and acoustic modes derives a wave equation for enthalpy pulsations, capturing both pseudo-sonic and acoustic oscillations. The model improves computational accuracy while accounting for turbulence, rotor interference, and impedance effects.

Finally, the fifth chapter explores Taylor–Couette flow with an axial temperature gradient through experiments and simulations. It identifies a specific regime where alternating Taylor vortices and thermal convection significantly enhance heat transfer. Numerical analysis confirms that this regime achieves a much higher Nusselt number with lower power input. These findings provide a promising strategy for energy-efficient thermal transport in rotating systems.

By bringing together these diverse yet interconnected topics, this book aims to provide researchers, engineers, and students with an in-depth exploration of vortex dynamics from multiple perspectives. The contributions presented here reflect the latest advancements in the field and offer a foundation for future studies on the fundamental and applied aspects of vortex behavior.

The editor sincerely thanks the authors for their invaluable contributions and Ms. Tea Jelaca, Publishing Process Manager, for her precise editorial support. The editor hopes this book serves as a valuable resource for those interested in the ever-evolving study of vortex dynamics.

Naoto Ohmura
Graduate School of Engineering,
Kobe University,
Kobe, Japan

Section 1

Theoretical Approaches

Chapter 1

Proton Properties from Nested Surface Vortices

Steven Verrall, Kelly S. Verrall, Andrew Kaminsky,
Isaac Ozolins, Emily Friederick, Andrew Otto, Ivan Ngian,
Reagen McCormick and Pearl Scallon

Abstract

A nested surface vortex structure may be used to explain several properties of free or chemically bound protons. The circular Unruh and zitterbewegung effects are combined to show that it is plausible for the mass of an unobserved ground-state proton to exist on a spherical surface. Such a model is consistent with general relativity. The charge of an unobserved ground-state proton is assumed to exist on two massless oppositely charged shells well outside that of its mass sphere. These two charge shells are assumed to exist on the two surfaces of a spindle torus. This spindle torus structure offers geometric explanations for proton isospin, *g*-factor, and charge radius. This geometric model involves mathematics typically encountered by undergraduate physics and chemistry students. Upon interaction with other particles, this ground-state proton model transforms into the valence quarks, gluon flux tubes, and initial sea quarks of the standard quantum chromodynamics model.

Keywords: quantum vortex, zitterbewegung fermion, circular Unruh effect, general relativity, intrinsic charm quarks, proton *g*-factor, proton charge radius

1. Introduction

Primordial nucleosynthesis formed the first atomic nuclei. This process ended about 20 minutes after the Big Bang. The first stars and galaxies formed hundreds of millions of years later. During the time between these processes, about 75% of the mass of elemental matter was in the form of neutronless hydrogen-1 in its ionized, atomic, and molecular forms. Ionized hydrogen-1 is a bare proton. The nucleus of atomic hydrogen-1 is also a bare proton. Molecular hydrogen-1 consists of two protons chemically bound by an electron cloud.

A completely accurate proton model has remained elusive since at least 1917 when Ernest Rutherford first provided experimental evidence that all atoms contain protons. If the proton's internal structure is not fully understood, humanity may lack fundamental insights into the nature of matter. On the energy scale where matter forms the building blocks of earthly life, protons appear to precisely maintain several

IntechOpen

important parameters. These include net charge, rms charge radius, mass, magnetic moment, spin, isospin, and parity.

In biological systems, almost all hydrogen is in the form of either chemically bound protons or ionized protons. Magnetic resonance imaging (MRI) is a key diagnostic tool in modern medicine because proton magnetic moment is unaffected by chemical binding. Medical MRI relies on each proton's magnetic moment precisely resonating with radio-frequency waves to emit coherent radiation with compact direction, frequency, and phase.

Quantum field theory (QFT) is the foundation of the Standard Model of particle physics [1]. The Standard Model is not completely explained because several parameters must be experimentally determined. QFT applies operators to create and annihilate particles [2, 3]. This circumvents potential physical mechanisms that create and annihilate mass and charge.

This chapter summarizes a proposed mechanism where quantum networks of interfering virtual vacuum momenta continually regenerate the mass and charge of each free or chemically bound ground-state proton. This is called the ground-state quantum vortex (GSQV) proton model [4, 5]. In this model, it is assumed that one real spin-1 photon splits into two virtual circularly polarized spin-half photon vortices during proton-antiproton pair production. An antiproton contains antimatter in the form of antiquarks. Antimatter is of identical mass and opposite charge to matter. Matter and antimatter readily annihilate each other.

All elemental matter consists of spin-half particles. In the GSQV proton model, the mass-energy of a free or chemically bound spin-half proton is proposed to be generated by the toroidal revolution of a virtual photon [4]. The initial confinement of mass-energy may be obtained by combining the zitterbewegung and circular Unruh effects [4]. The toroidally revolving virtual photon is proposed to be circularly polarized. This results in a virtual poloidal vortex component that Reference [4] associates with isospin and assumes generates charge. Reference [5] proposes a mechanism where twin virtual poloidal circulations generate and maintain a GSQV proton's charge.

The GSQV proton model adds to QFT without replacing any of its long-established aspects. It supports the fundamental validity of quantum chromodynamics (QCD) [6, 7]. The GSQV proton model seamlessly merges with chiral effective field theory (EFT) [8] and lattice QCD [9, 10] at higher energies. A free or chemically bound proton, in its lowest energy (ground) state, is modeled as a completely coherent self-synchronizing vortex structure. The quantum vortex structure is formed from toroidally and poloidally circulating virtual fields in the form of standing waves. These virtual fields arise from the real electromagnetic fields of real photons, which presumably formed the first protons in the early universe, and standing waves of virtual quantum vacuum fields. The precise nature of the quantum vacuum remains an unsolved problem. However, vacuum energy must manifest in a charge-neutral way, and standing vacuum waves explain the experimentally supported Casimir effect [11].

The GSQV proton model is depicted in **Figure 1**. Proton mass-energy is concentrated in the relatively small blue central sphere, which Reference [4] calls a zitterbewegung fermion. This finding is summarized in Sections 2 and 4. The surrounding charge structures are massless and are proposed by Reference [5] to be formed from standing vacuum waves. This finding is summarized in Section 5. At an unspecified energy, minimally above the ground state, Reference [4] proposes that the GSQV proton transforms into the quarks and gluons of established QCD theory. This is depicted in **Figure 2** and summarized in Section 3. Therefore, at higher

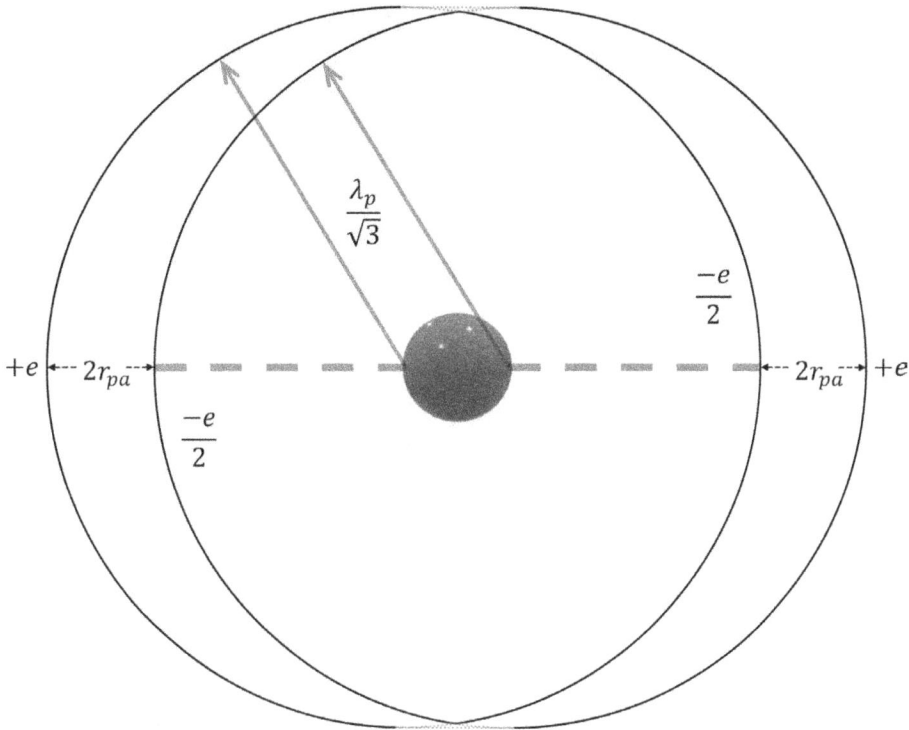

Figure 1.
The GSQV proton model consists of a central zitterbewegung fermion (blue sphere), of radius $r_{pz} = \frac{\lambda_p}{4\pi}$, orbited by four massless charge arcs. The outer two charge arcs are quantized to the value of the elementary charge, $+e$. The inner two charge arcs are quantized to the value of negative half the elementary charge, $-e/2$. Each charge arc orbits with its equator always moving at light speed, c. The inner and outer charge arcs occasionally align with the separation $2r_{pa}$. The distance between the two equatorial points of the inner charge arcs is always proton Compton wavelength, λ_p. Therefore, $r_{pa} = \lambda_p \left(\frac{1}{\sqrt{3}} - \frac{1}{2} \right)$, which is slightly less than r_{pz}. The red dots represent uncharged polar regions where virtual poloidal flow is split into two equal magnitudes.

energies, there should be no conflict with the established chiral EFT [8] and lattice QCD [9, 10] models. The GSQV proton model may help resolve recently discovered discrepancies occurring at the lowest energies [12–15]. Section 6 summarizes the Reference [5] finding that up, charm, and top quark charge appears to depend only on Planck charge, the proportionate area of a GSQV proton's polar charge-exclusion zone, and π. The charge-exclusion zones are indicated by the red dots in **Figure 1**. Section 7 summarizes how Reference [5] applies the GSQV proton model to calculate properties of intrinsic charm quarks. Section 8 summarizes how Reference [5] models the proton magnetic moment. Section 9 summarizes how Reference [5] calculates proton charge radii statistically consistent with the most accurate experimental estimates [16–18].

When in its ground state, Reference [5] proposes that proton charge and mass are coupled and continually regenerate each other. For each proton charge arc, this coupling may be represented in the form of a virtual optimal Möbius band [5, 19–22]. Reference [5] proposes that this implies the geometry of a GSQV proton is optimal, which may explain why free or chemically bound protons do not decay. In **Figure 1**, λ_p is proton Compton wavelength. Each proton charge arc is assumed to be regenerated by half a poloidal turn, at radius $R_p = \lambda_p/\sqrt{3}$, each zitterbewegung cycle. Reference [4]

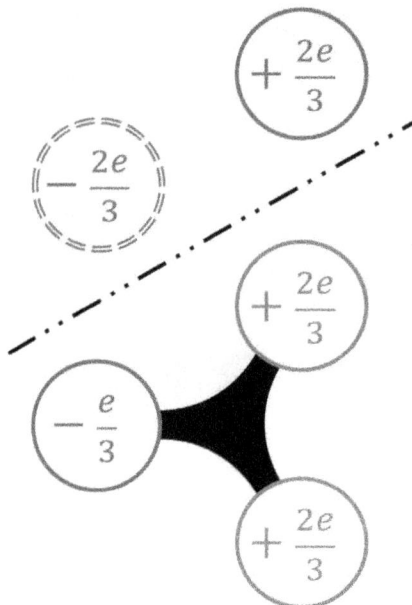

Figure 2.
Minimally excited free or chemically bound proton. Each circle represents a quark. Dashed outline represents an anticolor. The three quarks in the right-hand column transition from the two outer +e charge arcs of the GSQV proton. These momentarily form a color-neutral quark triplet. The two quarks in the left-hand column transition from the two inner −e/2 charge arcs of the GSQV proton. These momentarily form a color-neutral quark doublet. The three quarks below the dashed line connect via gluon flux tubes to form the proton's color-neutral valence quark triplet. The two quarks above the dashed line are unbound intrinsic sea quarks.

associates the half-poloidal turn with quantum mechanical isospin. The $\sqrt{3}$ factor is the aspect ratio of the optimal Möbius band and is applied extensively in Sections 5–9.

Since protons are spin-half particles, each zitterbewegung cycle involves two revolutions of the central zitterbewegung fermion [4, 5]. A proton zitterbewegung cycle may be characterized by Compton wavelength, λ_p, and Compton frequency, $f_p = c/\lambda_p$, where c is the speed of light.

2. Proton mass as quantized circular Unruh energy

References [4, 5] propose that the mass-energy of a free or chemically bound low-energy proton is equivalent to confined quantized circular Unruh energy [23–27]. The Unruh effect is known to be fundamentally local [28]. This circular Unruh energy is generated by the light-speed internal circulation of what Reference [4] calls a zitterbewegung fermion. To external observers, a zitterbewegung fermion consists of a uniformly distributed ensemble of light-speed circulations of point-like objects on the surface of a sphere of radius

$$r_{pz} = \frac{\hbar}{2m_p c} = \frac{\lambda_p}{4\pi},\tag{1}$$

where $\hbar$ is Planck's reduced constant and m_p is proton mass. Note that

$$\lambda_p = \frac{h}{m_p c} \quad \text{and} \quad \hbar = \frac{h}{2\pi}, \tag{2}$$

where h is Planck's constant. In **Figure 1**, the proton zitterbewegung fermion is depicted as a blue sphere. Its radius, r_{pz}, is clearly much smaller than the proton charge radius. Reference [4] proposes that each ensemble member of a zitterbewegung fermion is entangled with the rest of the ensemble.

In QFT, it is well established that any acceleration causes an increase in vacuum energy. This increase in vacuum energy can be described as Unruh temperature. Circular motion is caused by centripetal acceleration, which is associated with circular Unruh energy, T_{circ} [23]. To inertial observers, a free or chemically bound proton's zitterbewegung fermion consists of an ensemble of point-like objects with zitterbewegung acceleration [4],

$$a_{pz} = \frac{c^2}{r_{pz}} = \frac{2m_p c^3}{\hbar}. \tag{3}$$

Reference [4] shows that for a zitterbewegung fermion,

$$T_{\text{circ}} = \frac{\hbar a_{pz}}{4\sqrt{3}ck_B} = \frac{m_p c^2}{2\sqrt{3}k_B}, \tag{4}$$

where k_B is the Boltzmann constant. This implies that proton mass-energy,

$$E_p = m_p c^2 = 2\sqrt{3}k_B T_{\text{circ}}. \tag{5}$$

The median energy of a particle in a thermal bath at temperature, T, is given by $3.50302 k_B T$ [29]. This appears to imply that proton mass-energy, E_p, is approximately $2\sqrt{3}/3.50302 \approx 98.9\%$ of the median thermal excess vacuum energy due to internal zitterbewegung acceleration. References [5, 29] propose that the median energy is the most physically meaningful thermal spectrum peak. Reference [5] proposes that the approximately 1.1% discrepancy between E_p and the median thermal excess vacuum energy may explain the cosmic origin of quark masses.

The distance between the equatorial points of the two inner $-e/2$ charge arcs, depicted in **Figure 1**, is set to the proton Compton wavelength, λ_p. References [4, 5] propose that vacuum standing waves exist between the equatorial points of the inner $-e/2$ charge arcs. Section 5 summarizes this proposed phenomenon. Since these standing waves precisely match the proton's Compton wavelength, λ_p, Reference [4] proposes that they trigger the quantization of the Unruh energy of the central zitterbewegung fermion.

3. Proton quark formation

As proposed in Reference [4], a minimally excited GSQV proton may couple with the Higgs field and transfer mass-energy from the GSQV nucleon's central zitterbewegung fermion to its system of revolving formerly massless charge arcs. References [4, 5] propose that the $+2e$ charge of the outer GSQV proton charge arcs evenly divides into three up quarks of charge $+2e/3$. As proposed in Reference [5], it is likely that the two $+e$ charge arcs merge into a $+2e$ charge surface [4] before decomposing into three up quarks of charge $+2e/3$.

The structure shown in **Figure 1** rotates toroidally about the GSQV proton axis on surfaces of revolution. Apart from the flat polar caps, the rest of the toroidally rotating structure exists on the two surfaces of a spindle torus. The inner surface is called a lemon and the outer surface is called an apple. Further details can be found in Reference [4].

Quarks have long been known to be spin-half particles. Therefore, according to the Pauli exclusion principle, these three up quarks cannot be in the same quantum state. QCD invokes three fundamental color charge types to adhere to the Pauli exclusion principle. The right-hand column in **Figure 2** represents the outer GSQV proton shell transforming into a blue, green, and red color-neutral up-quark triplet.

Since a QCD proton is color neutral, and the GSQV proton's outer shell transforms into a color-neutral quark group, the GSQV proton's inner shell must also transform into a color-neutral quark group. At energies slightly above the proton's ground state, Reference [5] proposes that the two $-e/2$ inner charge arcs merge into a $-e$ charge shell [4] and divide into a down ($-e/3$) and antiup ($-2e/3$) quark pair. To maintain color neutrality, these two quarks must be a color-anticolor pair. The left-hand column in **Figure 2** represents the inner GSQV proton shell transforming into a blue down quark and an antiblue (shown as dashed blue) antiup quark. However, any color-anticolor combination is possible.

It is important to remember that color charge cycles extremely rapidly—to the degree that the color charge of each valence quark is effectively the superposition of all three colors and the color charge of each sea quark is the superposition of all three colors and all three anticolors.

Reference [5] proposes that the curvature of the charge shells prevents the initial formation of gluon flux tubes between quarks on the same shell. Initially, two gluon flux tubes are proposed to form between inner and outer valence quarks. Each of the two outer valence quarks is assumed to connect to the same inner valence quark *via* a gluon flux tube. This initial double gluon flux tube structure is then proposed to immediately transform into the three-way symmetric gluon flux tube structure described by the Standard Model of particle physics [30].

4. Ground-state strong force and gravitation equivalence

Reference [4] models the interior of the GSQV proton's zitterbewegung fermion as flat Minkowski spacetime. Excess vacuum energy, equaling proton mass-energy, is assumed contained inside the zitterbewegung fermion's spherical shell.

In flat Minkowski spacetime, trapped energy will exert pressure, $p = U/3V$, where U is trapped energy and V is volume. For the GSQV proton's spherical zitterbewegung fermion, $U = m_p c^2$ and $V = \frac{4}{3}\pi r_{pz}^3$. Therefore

$$p = \frac{U}{3V} = \frac{m_p c^2}{4\pi r_{pz}^3}. \tag{6}$$

The total outward force exerted on the zitterbewegung fermion's surface, $F_{\text{out}} = pA$, where $A = 4\pi r_{pz}^2$. Therefore

$$F_{\text{out}} = \frac{m_p c^2}{r_{pz}} = m_p a_{pz}, \tag{7}$$

where a_{pz} is zitterbewegung centripetal acceleration defined by Eq. (3). For the zitterbewegung fermion to be stable, there must be a balancing total inward force of value $F_{\text{in}} = F_{\text{out}}$. In the GSQV proton model, F_{in} plays the role of the strong force needed to confine proton mass-energy.

One of the foundational assumptions, used to derive the curved spacetime feature of general relativity, is the equivalence of gravitational and inertial mass. Inertial mass is itself equivalent to confined energy. It will therefore be assumed that the spacetime curvature of general relativity causes the total inward force, F_{in}, needed to stabilize the zitterbewegung spherical shell. This is equivalent to inward gravitational pressure,

$$p_g = \frac{m_p c^2}{4\pi r_{pz}^3}. \tag{8}$$

Reference [4] supposes that the point-like objects described in Section 2 are actually spheres of radius $\sqrt{2}\, l_P$, where $l_P \approx 1.616 \times 10^{-35}$m is the Planck length. Each circling sphere will have a cross-sectional area $2\pi l_P^2$ and always be located on the zitterbewegung spherical shell. The inward gravitational force, F_g, due to inward gravitational pressure, p_g, on each circling sphere, will be assumed to be a cross-sectional area, $2\pi l_P^2$, multiplied by inward gravitational pressure:

$$F_g = 2\pi l_P^2 p_g = \frac{m_p c^2 l_P^2}{2r_{pz}^3} = \frac{G m_p m_P l_P}{2r_{pz}^3} = \frac{G m_p^2}{r_{pz}^2}, \tag{9}$$

since $c^2 = G m_P / l_P$ and $m_P l_P / r_{pz} = \hbar / c r_{pz} = 2 m_p$, where G is the universal gravitational constant and m_P is the Planck mass. Each circling sphere therefore experiences a force, due to curved spacetime, as if it were a point-like proton mass, m_p, in the Newtonian gravitational field of another point-like proton mass at the center of the zitterbewegung fermion.

The same effect would occur if the point-like proton mass, at the center of the zitterbewegung fermion, was uniformly distributed on a thin spherical shell of radius $\leq r_{pz}$. Such a structure would exert no interior gravitational field, and therefore no interior spacetime curvature, but would exert the same external gravitational field as that of a point-like proton mass. It is therefore possible to model the zitterbewegung fermion's interior as flat Minkowski spacetime, which was assumed when deriving Eq. (6).

Proton mass uniformly distributed on a thin spherical shell, of radius r_{pz}, is equivalent to the mass distribution of the superposition of all zitterbewegung fermion ensemble members. Therefore, each ensemble member is effectively gravitationally entangled with the rest of the ensemble.

In particle physics, the strong and weak forces are often referred to as interactions. This is because they can be mathematically described as an exchange of virtual particles. The same is true of the mathematical description of electromagnetism in particle physics. The terms "force" and "interaction" can generally be interchanged. The term CP-symmetry refers to the combination of charge conjugation symmetry and parity symmetry. For more than 50 years, it has been established both experimentally and theoretically that the weak force can violate CP-symmetry.

According to long-established QCD theory, it may also be possible for the strong force to violate CP-symmetry. However, no experiment involving only the strong force has violated CP-symmetry. As QCD theory provides no fundamental reason for CP-symmetry to be conserved, this is known as the strong CP problem. No experiment has shown gravity to violate CP-symmetry. The connection made in Reference [4], and this section, between the strong force and gravity may imply that the strong force is CP-symmetric because gravity is CP-symmetric.

5. Proton charge arcs as quantum networks

In **Figure 1**, each inner charge arc shares its center with an outer charge arc. Reference [5] proposes that each inner-outer charge arc pair is generated by an ensemble of virtual photon standing waves with a uniform poloidal distribution. This finding is summarized in Section 6.

The charge arcs, proposed in Reference [5], are assumed to toroidally rotate about the proton axis and generate the charge surfaces proposed in Reference [4]. Reference [4] also proposes that an ensemble of standing waves, precisely matching proton Compton wavelength, λ_p, exists across the equatorial diameter of the inner charge surface. This implies that a single standing wave, of wavelength λ_p, exists between the equatorial points of the proton inner charge arcs as shown in **Figure 1**.

Reference [5] proposes that this equatorial standing wave is actually the superposition of the fundamental and second harmonics. Inside an infinite square well, the fundamental harmonic is a half wavelength and the second harmonic is a full wavelength. Reference [5] proposes that each poloidal ensemble of virtual photon standing waves interferes with the equatorial fundamental harmonic to generate the equatorial second harmonic.

Note that photon momentum is equal to Planck's constant, h, divided by wavelength. Since the inner-arc equatorial points are λ_p apart, the fundamental harmonic will be of wavelength $2\lambda_p$ and momentum $h/2\lambda_p$. Reference [5] assumes that the poloidal ensembles of virtual standing waves consist of fundamental harmonics of wavelength $2\lambda_p/\sqrt{3}$ and momentum $\sqrt{3}h/2\lambda_p$. The $\sqrt{3}$ factor is due to assumed charge and mass coupling *via* a virtual optimal Möbius band geometry. The rms value of the sum of interfering momenta $h/2\lambda_p$ and $\sqrt{3}h/2\lambda_p$ is given by

$$p_{\text{rms}} = \sqrt{\left(\frac{h}{2\lambda_p}\right)^2 + \left(\frac{\sqrt{3}h}{2\lambda_p}\right)^2} = \frac{h}{2\lambda_p}\sqrt{1^2 + \left(\sqrt{3}\right)^2} = \frac{h}{\lambda_p}. \tag{10}$$

This is the momentum of the second harmonic, of wavelength λ_p, presumed to exist between the equatorial points of the GSQV proton's inner charge arcs. This key assumption was used to develop the original GSQV proton model [4]. Reference [5] therefore proposes that the second equatorial harmonic is quantized as the rms value of the interference between the fundamental equatorial harmonic and the poloidal distributions of fundamental harmonics.

6. Charge arc generation and proton charge-exclusion zone

Reference [5] proposes that the charge structure of a GSQV proton consists of a pair of $+e$ charge arcs and a pair of $-e/2$ charge arcs. These are shown in **Figure 1**.

Each charge arc's equatorial point is farthermost from the proton axis. The equatorial point on each charge arc is proposed to toroidally revolve about the proton axis at light speed. It is this charge motion that generates the proton magnetic moment.

Reference [5] proposes that a standing wave ensemble connects each charge arc to its center and defines proton zitterbewegung inertial power (ZIP) as

$$P_p = E_p f_p, \tag{11}$$

where $E_p = hf_p$ is proton mass-energy and f_p is proton Compton frequency. Therefore

$$P_p = hf_p^2. \tag{12}$$

Reference [5] proposes that each charge arc is continually reflecting and accelerating virtual photons. Acceleration density per poloidal radian is assumed divided evenly among four charge arcs:

$$\frac{da_{\text{pol}}}{d\phi} = \frac{1}{4} R_p \omega_p^2, \tag{13}$$

where a_{pol} denotes the magnitude of poloidal centripetal acceleration, ϕ denotes latitude, and proton Compton angular frequency, $\omega_p = 2\pi f_p$, implies the key assumption that GSQV proton mass and charge regenerate at the same rate.

Reference [5] shows that Eq. (13) may be integrated to obtain

$$a_{\text{pol}} = \frac{1}{4} R_p \omega_p^2 \Delta\phi_{\text{arc}}, \tag{14}$$

where $\Delta\phi_{\text{arc}}$ is the angular extent of the charge arc. This poloidal acceleration is assumed to generate the four charge arcs of a GSQV proton. The quantity q_{arc} is defined by Reference [5] as the charge of each arc. Reference [5] applies the classical Larmor formula to describe the amount of ZIP, P_{pol}, which generates the four charge arcs:

$$P_{\text{pol}} = 4 \frac{q_{\text{arc}}^2 a_{\text{pol}}^2}{6\pi\varepsilon_0 c^3}, \tag{15}$$

where ε_0 is the electric permittivity of free space.

References [4, 5] define Q_{pex} as the uncharged proportion of the GSQV proton's outer surface, which resembles a flat cap at each pole. It follows that $\left(1 - Q_{pex}\right)$ represents the charged proportion of the GSQV proton's outer surface. For the outer charge arcs, it is reasonable to assume [5] that

$$P_{\text{pol}} = \left(1 - Q_{pex}\right)P_p = \left(1 - Q_{pex}\right)hf_p^2. \tag{16}$$

Substituting $R_p = \lambda_p/\sqrt{3}$ and $\Delta\phi_{\text{arc}} = \pi$ into Eq. (14) yields

$$a_{\text{pol}} = \frac{\pi\lambda_p}{4\sqrt{3}} \omega_p^2. \tag{17}$$

Reference [5] shows that

$$\lambda_p \omega_p^2 = 4\pi^2 c f_p. \tag{18}$$

Therefore

$$a_{\text{pol}} = \frac{\pi^3}{\sqrt{3}} c f_p. \tag{19}$$

Squaring yields

$$a_{\text{pol}}^2 = \frac{\pi^6}{3} c^2 f_p^2. \tag{20}$$

For a proton outer charge arc, $q_{\text{arc}} = e$. Substituting Eq. (20) into Eq. (15) yields

$$P_{\text{pol}} = \frac{2\pi^5 e^2 f_p^2}{9\varepsilon_0 c}. \tag{21}$$

Substituting Eq. (16) into Eq. (21) yields

$$\left(1 - Q_{pex}\right)h = \frac{2\pi^5 e^2}{9\varepsilon_0 c}. \tag{22}$$

Rearranging,

$$e^2 = \frac{9\varepsilon_0 hc}{2\pi^5}\left(1 - Q_{pex}\right). \tag{23}$$

The asymptotic low-energy value of the fine-structure constant [1, 2],

$$\alpha = \frac{e^2}{2\varepsilon_0 hc}. \tag{24}$$

Substituting Eq. (23) into Eq. (24) yields

$$\alpha = \frac{9}{4\pi^5}\left(1 - Q_{pex}\right). \tag{25}$$

Presuming Eq. (25) to be exact, the 2022 Committee on Data of the International Science Council (CODATA) value [31],

$$\alpha = 7.2973525643(11) \times 10^{-3}, \tag{26}$$

can be used to calculate

$$Q_{pex} = 1 - \frac{4}{9}\pi^5\alpha = 1 - \alpha\pi^5\left(\frac{q_u}{e}\right)^2 = 7.49620822(15) \times 10^{-3}, \tag{27}$$

where $q_u = 2e/3$ is the charge of an up, charm, or top quark. Note that this Q_{pex} value is only 2.1% larger than that estimated in Reference [4]. This numerically supports the assumption $R_p = \lambda_p/\sqrt{3}$, which is tied to the assumption that charge and mass are coupled *via* a virtual optimal Möbius band geometry [5].

Planck charge, q_P, may now be written in terms of e and α as

$$q_P^2 = \frac{e^2}{\alpha} = \frac{4\pi^5 e^2}{9\left(1 - Q_{pex}\right)}, \tag{28}$$

where e^2 was divided by Eq. (25). Taking the square root,

$$q_P = \frac{2e}{3}\sqrt{\frac{\pi^5}{\left(1 - Q_{pex}\right)}} = q_u \sqrt{\pi^5/\left(1 - Q_{pex}\right)}. \tag{29}$$

Rearranging,

$$q_u = q_P \sqrt{\left(1 - Q_{pex}\right)/\pi^5}, \tag{30}$$

Surprisingly up, charm, and top quark charge appears to depend only on Planck charge, the geometric constant π, and the charged proportion of the GSQV proton's outer surface, $\left(1 - Q_{pex}\right)$. It follows that

$$q_d = \frac{q_P}{2}\sqrt{\left(1 - Q_{pex}\right)/\pi^5}, \tag{31}$$

where $q_d = e/3$ is the charge of a down, strange, or bottom quark.

7. Intrinsic charm quark formation

Reference [5] proposes that the intrinsic up-antiup (u$\bar{\text{u}}$) virtual sea quark pair, depicted in **Figure 2**, may occasionally transform into a charm quark, c, and an anticharm quark, $\bar{\text{c}}$. This c$\bar{\text{c}}$ quark pair may momentarily form bound states, such as |uudc$\bar{\text{c}}$>, with the proton's three valence quarks [32]. Reference [5] proposes that twin poloidal revolutions, phase-locked with each other, continually regenerate charge. Reference [4] associates poloidal revolution with both isospin and charge generation. The two vectors shown in **Figure 1** always have the same poloidal angle, relative to the equator, as they circulate. These two poloidal vectors occasionally overlap. The overlaps are indicated by the heavy dashed lines in **Figure 1**. Two such overlaps will occur in each poloidal cycle, which is the same time duration as a zitterbewegung cycle because charge and mass are assumed to regenerate at the same rate.

Reference [5] proposes that each of these overlapping lengths exists momentarily as a standing wave of vacuum energy. If each wave is a fundamental harmonic, it will be of wavelength

$$2\left(\frac{\lambda_p}{\sqrt{3}} - 2r_{pa}\right) = 2\lambda_p\left(1 - \frac{1}{\sqrt{3}}\right). \tag{32}$$

This is shorter than the wavelength of the fundamental harmonic ensemble regenerating the charge arcs, $2R_p = 2\lambda_p/\sqrt{3}$, explained in Section 5 and Reference [5]. The proportion of shortening is clearly

$$\sqrt{3}\left(1 - \frac{1}{\sqrt{3}}\right) = \sqrt{3} - 1. \tag{33}$$

Reference [5] proposes that this shortening is additive and equivalent to proton mass-energy momentarily increasing by the factor $\left(\sqrt{3} - 1\right)^{-1}$. This is evaluated as

$$\frac{m_p}{\sqrt{3} - 1} \approx 1281.70\,\text{MeV}/c^2, \tag{34}$$

which is slightly more energy than that needed to generate either a charm or anticharm quark "running" mass [17]. The momentary proton mass-energy increase should appear to be twice this amount because there are two momentary overlaps in each poloidal cycle. This process therefore provides more than enough additional energy to generate both the charm and anticharm quark "running" masses [17]. Since this charm-anticharm quark pair forms from an up-antiup quark pair, Reference [5] goes on to calculate charm quark "running" mass as about $1279\,\text{MeV}/c^2$. This value is well inside the 2022 recommended Particle Data Group range: $m_c = 1270 \pm 20\,\text{MeV}/c^2$ [17].

Even though the twin charge-regenerating vectors are poloidally phase-locked, they are not toroidally synchronized. Reference [5] supposes that the twin charge-regenerating vectors become entangled, and generate additional mass-energy, whenever both vector tips are sufficiently close to the zitterbewegung equator. Sufficiently close is assumed to be closer than the zitterbewegung radius [4], r_{cz}, of the charm quark "running" mass, m_c [17].

During this entanglement, the maximum angular separation, θ_{ent}, between the zitterbewegung equator and each R_p vector will be

$$\theta_{ent} = \sin^{-1}\left(\frac{r_{cz}}{R_p}\right) = \sin^{-1}\left(\frac{\sqrt{3}\,m_p}{4\pi m_c}\right). \tag{35}$$

Reference [5] shows that all possible entanglements will occur on an angular area that is

$$\frac{(1 - \cos\theta_{ent})}{2} = \frac{1}{2}\left(1 - \sqrt{1 - \frac{3m_p^2}{16\pi^2 m_c^2}}\right) \approx 0.26\% \tag{36}$$

of the total spherical angular area. This implies the $|uudc\bar{c}>$ bound state is present about 0.26% of the time. Reference [5] calculates the fraction of proton momentum carried by intrinsic charm quarks as

$$\frac{m_c}{m_p}(1 - \cos\theta_{ent}) = \frac{m_c}{m_p} - \sqrt{\left(\frac{m_c}{m_p}\right)^2 - \frac{3}{16\pi^2}} = 0.704(11)\%. \tag{37}$$

The uncertainty has been propagated from the 2022 recommended Particle Data Group range: $m_c = 1270 \pm 20 \, \text{MeV}/c^2$ [17]. Reference [32] reports an experimental estimate $(0.62 \pm 0.28)\%$. Our calculation, displayed as Eq. (37), is consistent with this experimental range and about 25 times more precise.

8. Proton magnetic moment from charge arcs

Reference [5] proposes the following equation for proton magnetic moment:

$$\mu_p = \left(\frac{2V_{po}}{\left(1+\delta_p\right)\left(1-Q_{pex}\right)\left(R_p + r_{pa}\right)^2} - \frac{V_{pi}}{\left(1-\delta_p\right)\left(R_p - r_{pa}\right)^2} \right) \frac{\mu_N}{4\pi r_{pz}}, \qquad (38)$$

where μ_N is the nuclear magneton unit typically used to express magnetic moments of atomic nuclei. The quantity V_{pi} is the lemon volume formed by toroidally rotating the GSQV proton's two inner charge arcs about the GSQV proton axis. The quantity V_{po} is the volume formed by toroidally rotating the GSQV proton's two outer charge arcs, with flat end caps, about the GSQV proton axis. Applying calculus results derived in the Appendix of Reference [4],

$$V_{pi} = \frac{4}{3}\pi R_p^3 \left[\sin^3 \phi_{pl} - \frac{3}{4} \cos \phi_{pl} \left(2\phi_{pl} - \sin 2\phi_{pl} \right) \right] \qquad (39)$$

and

$$V_{po} = \pi^2 R_p^2 \left(r_{pa} + \frac{4R_p}{3\pi} \right) + 2\pi r_{pa}^2 R_p, \qquad (40)$$

where

$$\phi_{pl} = \cos^{-1}\left(\frac{r_{pa}}{R_p}\right) = \cos^{-1}\left(1 - \frac{\sqrt{3}}{2}\right) \qquad (41)$$

and

$$r_{pa} = R_p - \frac{\lambda_p}{2} = \lambda_p \left(\frac{1}{\sqrt{3}} - \frac{1}{2} \right). \qquad (42)$$

Substituting Eqs. (1) and (42) into Eq. (38) yields

$$\mu_p = \left(\frac{6V_{po}}{\left(1+\delta_p\right)\left(1-Q_{pex}\right)\left(4-\sqrt{3}\right)^2} - \frac{V_{pi}}{1-\delta_p} \right) \frac{4}{\lambda_p^3} \frac{\mu_N}{}. \qquad (43)$$

Noting $R_p = \lambda_p/\sqrt{3}$, it can be seen from Eqs. (39)–(42) that both V_{pi} and V_{po} are proportional to λ_p^3. A nucleus g-factor is twice the value of its magnetic moment when expressed with the nuclear magneton unit. Since both Q_{pex} and δ_p are unitless, Eq. (43) implies proton g-factor is independent of proton mass.

The quantity δ_p is a positive dimensionless parameter with a value less than 1/1000. In Eqs. (38) and (43), δ_p has the effect of perturbing the charge distribution slightly away from the equator on the outer charge arcs and slightly toward the equator on the inner charge arcs.

The reason for these slight perturbations may be explained subjectively as the self-interaction of the toroidally rotating charge arcs with their electromagnetic fields. However, it must be stressed that this proposed self-interaction mechanism subjectively applies classical electromagnetism to the interior of a quantum mechanical particle. This is beyond known physics. As such, δ_p is used as an adjustable parameter. While we do not calculate δ_p from first principles, Reference [5] shows that it can be calculated from a similar adjustable parameter defined for a GSQV neutron model, neutron mass, and the sum of the up and down quark masses.

A quadratic equation, in terms of δ_p, may be obtained by rearranging Eq. (43):

$$a\delta_p^2 + b\delta_p + c = 0, \qquad (44)$$

where

$$a = -\left(4 - \sqrt{3}\right)^2 \left(1 - Q_{pex}\right) \frac{\lambda_p^3 \mu_p}{4\mu_N}, \qquad (45)$$

$$b = \left(4 - \sqrt{3}\right)^2 \left(1 - Q_{pex}\right) V_{pi} + 6V_{po}, \qquad (46)$$

and

$$c = b - 12V_{po} - a. \qquad (47)$$

Applying the quadratic formula,

$$\delta_p = \frac{-b \pm \sqrt{b^2 - 4ac}}{2a} = 2.7353468(15) \times 10^{-4} \text{ and } \approx 2.64, \qquad (48)$$

where the 2022 CODATA value of μ_p, displayed as Eq. (49), is input to Eq. (45). Note that the 2022 CODATA uncertainties in α and μ_p contribute almost equally to the uncertainty in δ_p. The second solution is unphysical, so $\delta_p = 2.7353468(15) \times 10^{-4}$ provides a calculated proton magnetic moment, μ_p, as precise as the 2022 CODATA value [31]:

$$\mu_p = 2.79284734463(82) \, \mu_N. \qquad (49)$$

Reference [5] shows how $\delta_p = 2.7353468(15) \times 10^{-4}$ can be input into a GSQV neutron model that calculates a neutron magnetic moment about two orders of magnitude more precisely than the most accurate experiments.

9. Effective charge radius

Reference [18] experimentally estimated proton polar axial charge radius, $r_A = 0.73 \pm 0.17$ fm. Based on **Figure 1**, the GSQV proton's effective polar charge radius is

$$r_A \approx R_p = \frac{\lambda_p}{\sqrt{3}} \approx 0.763 \, \text{fm}, \tag{50}$$

which is well within the range reported by Reference [18].

Reference [4] originally developed the GSQV proton model by assuming a charge distribution that is the radial projection of two uniformly charged concentric spheres. Due to the tiny value of δ_p, this key assumption still accurately reflects the GSQV proton charge distribution implied by Eqs. (38) and (43). The polar charge-exclusion zones on the GSQV proton's outer charge surface provide the most significant deviation from this key assumption.

If a GSQV proton did not have charge-exclusion zones, radially projecting all charge out to R_{Eo} would yield a spherically symmetric charge distribution with electric potential

$$U(R_{Eo}) = \frac{ke}{R_{Eo}} \tag{51}$$

at distance R_{Eo} from the proton center, where k is Coulomb's constant. The GSQV proton model is approximately spherical. Therefore, electric potential on its outer surface, at average distance

$$r_s = \sqrt[3]{\frac{3}{4\pi} V_{po}} \tag{52}$$

from the proton center, may be approximated by

$$U(r_s) \approx \frac{ke}{r_s}. \tag{53}$$

This implies

$$\frac{U(r_s)}{ke} \approx \frac{1}{r_s}. \tag{54}$$

For a sphere,

$$\frac{rA}{V} = 3, \tag{55}$$

where r is radius, A is surface area, and V is volume. Since a perfect sphere has a minimum surface area to volume ratio for its size, rA/V should be greater than 3 for a perturbed sphere. The Appendix of Reference [4] shows how to calculate the surface area of volume V_{po}:

$$A_{po} = 2\pi^2 R_p \left(r_{pa} + \frac{2R_p}{\pi} \right) + 2\pi r_{pa}^2. \tag{56}$$

For the GSQV proton model of Section 8,

$$\frac{r_s A_{po}}{V_{po}} \approx 3.0059. \tag{57}$$

To a first approximation, combining Eqs. (54) and (57) yields

$$\frac{U(r_s)}{ke} \approx \frac{A_{po}}{3V_{po}}.$$

(58)

This implies that a change in charge volume would have about a third of the effect, on a muon or electron orbital, as a change in outer surface charge area. The GSQV proton outer surface charge area,

$$A_{\text{eff}} = A_{po}\left(1 - Q_{pex}\right).$$

(59)

This is assumed equivalent to

$$V_{\text{eff}} \approx V_{po}\left(1 - \frac{Q_{pex}}{3}\right),$$

(60)

which implies an effective charge radius,

$$r_{\text{eff}} = \sqrt[3]{\frac{3}{4\pi}V_{\text{eff}}} \approx \sqrt[3]{\frac{3}{4\pi}V_{po}\left(1 - \frac{Q_{pex}}{3}\right)} \approx 0.8409 \ \text{fm}.$$

(61)

This value is consistent with both the 2022 Particle Data Group recommended value, $r_p = 0.8409(4)\,\text{fm}$ [16, 17], and the 2022 CODATA recommended value, $r_p = 0.84075(64)\,\text{fm}$ [31].

10. Conclusions

This chapter summarizes our previously reported findings [4, 5] that a nested surface vortex structure can explain several properties of free or chemically bound protons. We call this the GSQV proton model. This geometric model can be visualized in the usual 3 spatial dimensions, with mathematics not beyond that typically encountered by undergraduate physics and chemistry students. Additional details can be found in References [4, 5]. Reference [5] includes a GSQV neutron model and proposes novel mechanisms that link proton and neutron properties.

Sections 2 and 4 summarized the Reference [4] finding that GSQV proton mass-energy may be concentrated in a relatively small central sphere called a zitterbewegung fermion. This aspect of the model is consistent with general relativity. Section 5 summarized the Reference [5] proposal that the surrounding massless charge structures are formed from standing vacuum waves. These charge structures take the form of arcs rotating about the proton axis. This rotation is set to light speed at the equator, and causes the proton charge to be distributed on the inner lemon and outer apple surfaces of a spindle torus. The polar dimples of the spindle torus are uncharged.

At higher energies, this model transforms into the valence quarks, gluon flux tubes, and initial sea quarks of the standard quantum chromodynamics model. This was summarized in Section 3. With the established chiral EFT and lattice QCD models, recently discovered discrepancies with experiment occur at the lowest energies. The GSQV proton model may therefore help resolve these discrepancies.

Section 6 summarized the Reference [5] finding that up, charm, and top quark charge depends only on Planck charge, the proportionate area of a GSQV proton's polar charge-exclusion zone, and π. Section 7 reviewed how the GSQV proton model can calculate properties of intrinsic charm quarks. Section 8 reviewed the proton magnetic moment calculation. Section 9 summarized the Reference [5] finding that effective proton charge radii are statistically consistent with the most accurate experimental estimates.

Acknowledgements

The authors thank Zhijun Jia for his advice on how to write an appropriate introduction and for his encouragement and endless enthusiasm. The authors also thank Kori Verrall for her helpful discussions and instruction on the optimal Möbius band. The authors also thank Dean Ju Kim and Associate Dean Gubbi Sudhakaran for their helpful technical discussions and key administrative support. Most of the original concepts were developed, while S.V. was partly supported by UWL Faculty Research Grant 23-01-SV and A.O. was supported by UWL URC grant #35F22. The CPC was funded by the authors.

Conflicts of interest

During the course of this research, Andrew Kaminsky graduated from the University of Wisconsin at La Crosse and was an employee at Benchmark and later PDA Engineering. These employment relationships had no influence on this research.

During the course of this research, Isaac Ozolins graduated from the University of Wisconsin at La Crosse and was an employee at ThermTech. This employment relationship had no influence on this research.

During the course of this research, Andrew Otto graduated from the University of Wisconsin at La Crosse and was an employee at St. Croix Health. This employment relationship had no influence on this research.

Nomenclature

A	sphere surface area
A_{eff}	proton outer surface charge area
A_{po}	proton outer surface area
a_{pol}	poloidal centripetal acceleration
a_{pz}	proton zitterbewegung acceleration
c	speed of light in vacuum $= 299792458\,\text{m/s}$
c	charm quark
$\bar{c}$	anticharm quark
CODATA	Committee on Data of the International Science Council
CP	charge parity
d	down quark
e	elementary charge $= 1.602176634 \times 10^{-19}\text{C}$
E_p	proton mass-energy

EFT	effective field theory
F_g	gravitational force
F_{in}	total inward force
F_{out}	total outward force
f_p	proton Compton frequency
G	universal gravitational constant
GSQV	ground-state quantum vortex
h	Planck constant $= 6.62607015 \times 10^{-34}$ J/Hz
$\hbar$	reduced Planck constant $= h/2\pi$
k	Coulomb constant
k_B	Boltzmann constant
l_P	Planck length
m_c	charm quark mass
m_P	Planck mass
m_p	proton mass
MRI	magnetic resonance imaging
$p_g = p$	proton gravitational pressure
P_p	proton zitterbewegung inertial power
P_{pol}	amount of P_p that regenerates the four charge arcs
p_{rms}	rms momentum
q_{arc}	arc charge
q_d	charge of a down, strange, or bottom quark
q_P	Planck charge
q_u	charge of an up, charm, or top quark
QCD	quantum chromodynamics
QFT	quantum field theory
Q_{pex}	uncharged proportion of GSQV proton outer surface
r	sphere radius
r_A	proton polar axial charge radius
r_{cz}	charm quark zitterbewegung radius
r_{eff}	proton effective charge radius
R_{Eo}	radius of spherically symmetric charge distribution
R_p	proton charge arc radius
r_p	proton rms charge radius
r_{pa}	half minimum equatorial separation between charge arcs
r_{pz}	proton zitterbewegung radius
r_s	radius of sphere with volume V_{po}
T	temperature
T_{circ}	temperature of circular Unruh energy
U	potential energy
u	up quark
$\bar{u}$	antiup quark
V	sphere volume
V_{eff}	proton effective charge volume
V_{pi}	proton lemon volume
V_{po}	proton outer volume
ZIP	zitterbewegung inertial power
α	fine-structure constant

δ_p	dimensionless fine-tuning parameter
ε_0	electric permittivity of free space
θ_{ent}	maximum angular separation
λ_p	proton Compton wavelength
μ_N	nuclear magneton
μ_p	proton magnetic moment
π	ratio of circle circumference to diameter
ϕ	poloidal latitude
ϕ_{pl}	maximum poloidal latitude of an inner charge arc
$\Delta\phi_{\text{arc}}$	angular extent of an outer charge arc
ω_p	proton Compton angular frequency

Author details

Steven Verrall[1*†], Kelly S. Verrall[2†], Andrew Kaminsky[1,3,4†], Isaac Ozolins[1,5†], Emily Friederick[1†], Andrew Otto[1,6†], Ivan Ngian[1†], Reagen McCormick[1†] and Pearl Scallon[1†]

1 University of Wisconsin at La Crosse, La Crosse, WI, USA

2 Independent Researcher, La Crosse, WI, USA

3 PDA Engineering, Burnsville, MN, USA

4 Benchmark, Winona, MN, USA

5 ThermTech Inc., Waukesha, WI, USA

6 St. Croix Health, St. Croix Falls, WI, USA

*Address all correspondence to: steven.verrall@gmail.com

† These authors contributed equally.

IntechOpen

References

[1] Oerter R. The Theory of Almost Everything: The Standard Model, the Unsung Triumph of Modern Physics. New York: Pi Press; 2006. 327 p

[2] Peskin M, Schroeder D. An Introduction to Quantum Field Theory. Boca Raton: CRC Press; 1995. 866 p. DOI: 10.1201/9780429503559

[3] Sebens CT. The fundamentality of fields. Synthese. 2022;**200**:1-28. DOI: 10.1007/s11229-022-03844-2

[4] Verrall SC, Atkins M, Kaminsky A, et al. Ground state quantum vortex proton model. Foundations of Physics. 2023;**53**. DOI: 10.1007/s10701-023-00669-y

[5] Verrall SC, Kaminsky A, Verrall KN, et al. Unbound low-energy nucleons as semiclassical quantum networks. 2024. DOI: 10.20944/preprints202405.0932.v1 [Preprint]

[6] Greensite J. An Introduction to the Confinement Problem. 2nd ed. Cham, Switzerland: Springer Nature; 2020. 271 p. DOI: 10.1007/978-3-030-51563-8

[7] Achenbach P, Adhikari D, Afanasev A, et al. The present and future of QCD. Nuclear Physics A. 2024; **1047**. DOI: 10.1016/j.nuclphysa.2024.122874

[8] Bernard V, Kaiser N, Ulf-G M. Chiral dynamics in nucleons and nuclei. International Journal of Modern Physics E: Nuclear Physics. 1995;**4**:193-344. DOI: 10.1142/S0218301395000092

[9] NPLQCD Collaboration. Magnetic moments of light nuclei from lattice quantum chromodynamics. Physical Review Letters. 2014;**113**. DOI: 10.1103/PhysRevLett.113.252001

[10] Bhattacharya S, Cichy K, Constantinou M, et al. Moments of proton GPDs from the OPE of nonlocal quark bilinears up to NNLO. Physical Review D. 2023;**108**. DOI: 10.1103/PhysRevD.108.014507

[11] Lamoreaux SK. The Casimir force: Background, experiments, and applications. Reports on Progress in Physics. 2005;**68**. DOI: 10.1088/0034-4885/68/1/R04

[12] Zheng X, Deur A, Kang H, et al. Measurement of the proton spin structure at long distances. Nature Physics. 2021;**17**:736-741. DOI: 10.1038/s41567-021-01198-z

[13] Deur A, Chen JP, Kuhn SE, et al. Experimental study of the behavior of the Bjorken sum at very low Q^2. Physics Letters B. 2022;**825**. DOI: 10.1016/j.physletb.2022.136878

[14] Li R, Sparveris N, Atac H, et al. Measured proton electromagnetic structure deviates from theoretical predictions. Nature. 2022;**611**:265-270. DOI: 10.1038/s41586-022-05248-1

[15] Ruth D, Zielinski R, Gu C, et al. Proton spin structure and generalized polarizabilities in the strong quantum chromodynamics regime. Nature Physics. 2022;**18**:1441-1446. DOI: 10.1038/s41567-022-01781-y

[16] Antognini A et al. Proton structure from the measurement of 2S-2P transition frequencies of muonic hydrogen. Science. 2013;**339**:417-420. DOI: 10.1126/science.1230016

[17] Workman RL et al. Review of particle physics. PTEP. 2022;**2022**. DOI: 10.1093/ptep/ptac097

[18] Cai T, Moore ML, Olivier A, et al. Measurement of the axial vector form factor from antineutrino-proton scattering. Nature. 2023;**614**:48-53. DOI: 10.1038/s41586-022-05478-3

[19] Wunderlich W. Über ein abwickelbares Möbiusband. Monatshefte für Mathematik. 1962;**66**:276-289. DOI: 10.1007/BF01299052

[20] Halpern B, Weaver C. Inverting a cylinder through isometric immersions and embeddings. Transactions of the American Mathematical Society. 1977; **230**:41-70. DOI: 10.1090/S0002-9947-1977-0474388-1

[21] Fuchs D, Tabachnikov S. Mathematical Omnibus: Thirty Lectures on Classic Mathematics. Vol. 46. Providence, RI, USA: American Mathematical Society; 2007. 465 p. DOI: 10.1090/mbk/046

[22] Schwartz RE. The optimal paper Moebius band. Annals of Mathematics. DOI: 10.48550/arXiv.2308.12641 [Preprint]

[23] Biermann S, Erne S, Gooding C, et al. Unruh and analogue Unruh temperatures for circular motion in 3+1 and 2+1 dimensions. Physical Review D. 2020;**102**. DOI: 10.1103/PhysRevD. 102.085006

[24] Lochan K, Ulbricht H, Vinante A, Goyal SK. Detecting acceleration-enhanced vacuum fluctuations with atoms inside a cavity. Physical Review Letters. 2020;**125**. DOI: 10.1103/ PhysRevLett.125.241301

[25] Zhang J, Yu H. Entanglement harvesting for Unruh-DeWitt detectors in circular motion. Physical Review D. 2020;**102**. DOI: 10.1103/PhysRevD. 102.065013

[26] Zhou Y, Hu J, Yu H. Entanglement dynamics for Unruh-DeWitt detectors interacting with massive scalar fields: The Unruh and anti-Unruh effects. Journal of High Energy Physics. 2021;**88**. DOI: 10.1007/JHEP09(2021)088

[27] Zhou Y, Hu J, Yu H. Steady-state entanglement for rotating Unruh-DeWitt detectors. Physical Review D. 2022;**106**. DOI: 10.1103/PhysRevD. 106.105028

[28] Anastopoulos C, Savvidou N. Coherences of accelerated detectors and the local character of the Unruh effect. Journal of Mathematical Physics. 2012; **53**. DOI: 10.1063/1.3679554

[29] Heald M. Where is the "Wien peak"? American Journal of Physics. 2003;**71**: 1322-1323. DOI: 10.1119/1.1604387

[30] Pvoh B, Rith K, Scholz C, et al. Particles and Nuclei: An Introduction to the Physical Concepts. 7th ed. Heidelberg, Germany: Springer; 2015. 458 p. DOI: 10.1007/978-3-662-46321-5

[31] Mohr P, Tiesinga E, Newell D, Taylor B. Codata Internationally Recommended 2022 Values of the Fundamental Physical Constants [Internet]. Gaithersburg, MD, USA: National Institute of Standards and Technology (NIST); 2024. Available from: https://physics.nist.gov/constants [Accessed: June 4, 2024]

[32] NNPDF Collaboration. Evidence for intrinsic charm quarks in the proton. Nature. 2022;**608**:483-487. DOI: 10.1038/s41586-022-04998-2

Chapter 2

About Bellman Principle and Solution Properties for Navier–Stokes Equations in the 3D Cauchy Problem

Vladimir I. Semenov

Abstract

Without belittling the achievements of many mathematicians in the studying of the Navier-Stokes equations, the real ways opened J. Leray and O.A. Ladyzhenskaya. The main goal of this work is to compare the smoothness property of a weak solution in the Cauchy problem after some moment if it is known solution regularity until this moment with the optimality property in the Bellman principle. Naturally, all these are connected with the existence problem of blow up solution in the Cauchy problem for Navier-Stokes equations in space attracting a lot of attention up to now. The smoothness control and controlling parameters can be varied. It is important to control the dissipation of kinetic energy to the fix moment or rate of change of kinetic energy square or the summability of velocity gradient to the fixed point in time and so on. There are possible other control parameters due to a weak solution.

Keywords: Navier-Stokes equations, blow up solution, regular solution, turbulent flow, dissipation of kinetic energy

1. Introduction

The well-known principle of dynamic programming (Bellman principle) postulates that whatever the state of the system at any step and the control at this step selected, subsequent controls should be selected optimal relative to the state to which the system will arrive at the end of this step. We would like to get something similar for the smoothness properties of a weak solution in the Cauchy problem for the Navier-Stokes equations in space, i.e., we would like to find out the conditions under which the existing regularity of the solution in a short time interval (see Refs. [1, 2]), and some important parameters (control parameters) of this solution at a certain point in time, while this smoothness is still preserved, it will be able to ensure the existence of a global smooth extension of this solution, or at least of a local smooth extension for some guaranteed longer time interval. Let us explain this. The most important characteristic of a fluid flow is kinetic energy at every point in time. If kinetic energy changing at the fixed point in time is not large, then we have no phenomenon collapse.

IntechOpen

Hence, we obtain the one from the controlling parameters. We can also control a rate of change of kinetic energy square. Here, the situation is the same. If at the initial time or the later point in time this rate of change is not large, then we have no phenomenon blow up again. Really, instead of the point time controlling, we can apply some new characteristics which are described as the mean value of some quantity over the fixed time interval. The result will be the same. This generality shows that comparison with Bellman principle is relevant.

In this way, the first steps were undertaken in Ref. [3] where numerical parameters were introduced as control parameters. Indirectly, they control the solution smoothness over a longer time interval. This point of view differs from the classical methods in studying the smoothness properties of weak solutions (see Refs. [1, 2, 4–6]) where the main tools are connected with embedding lemmas and multiplicative inequalities. In fact, embedding lemmas and multiplicative inequalities are the main tools up to now.

Here, another interesting aspect should be noted. It is related to the asymptotics of smooth solutions at infinity and integral identities for solenoid fields (see, for example, [7]). It may be a new tool for new apriori estimates.

2. Notations

We consider a motion of an ideal incompressible fluid and the simplest problem that is Cauchy problem in space ($n = 3$) which is described by equations:

$$\frac{\partial u}{\partial t} + \sum_{i=1}^{3} u_i \frac{\partial u}{\partial x_i} = \nu \Delta u - \nabla P, \quad div\, u = 0, \quad u(0,x) = \varphi(x), \tag{1}$$

where $u = u(t,x) = (u_1(t,x), u_2(t,x), u_3(t,x))$ is velocity vector. Symbols t and $x = (x_1, x_2, x_3)$ are time and spatial variables, respectively. A function $P = P(t,x)$ is pressure function and a vector field $\varphi(x) = (\varphi_1(x), \varphi_2(x), \varphi_3(x))$ is initial data; a constant ν is viscosity coefficient.

By symbols Δ and ∇, we denote Laplace operator and gradient operator on spatial variables, respectively. In particular, ∇u is Jacobi matrix on spatial variables.

Next, we apply the following notation:

$$|\varphi| = \sqrt{\sum_{i=1}^{3} \varphi_i^2}, \quad |\nabla \varphi| = \sqrt{\sum_{i=1}^{3} |\nabla \varphi_i|^2},$$

$$\|u(t, \cdot)\|_p^p = \int_{R^3} |u(t,x)|^p dx, \quad \|\nabla u(t, \cdot)\|_p^p = \int_{R^3} |\nabla u(t,x)|^p dx, \quad p > 1. \tag{2}$$

Lebesgue class $L_p(R^3)$ is defined as a set of vector fields in R^3 with finite norm $\|v\|_2$. In particular, and this is very important, for solutions u of problem (1), the kinetic energy at moment t is expressed by mean (energy integral) $\|u(t, \cdot)\|_2^2$ (see Refs. [1, 6]).

In addition, we suppose that initial data φ, as vector field $\varphi \in C_{6/5,3/2}^\infty(R^3)$, that is, φ is infinitely differentiable mapping. It belongs to Lebesgue class $L_{6/5}(R^3)$, and the first partial derivatives $\nabla \varphi \in L_{3/2}(R^3)$ and the rest derivatives belong to class $L_r(R^3)$ for any

$r > 1$ (class $C^\infty_{6/5,3/2}(R^3)$, its properties, and usefulness are described in Ref. [3]). In particular, it implies the following inclusions $\varphi \in L_2(R^3)$, $\nabla \varphi \in L_2(R^3)$ (see Ref. [3], Lemma 32). For us, this class is interesting only because for any fixed t solution $u(t, \cdot)$ of Cauchy problem (1) belongs to class $C^\infty_{6/5,3/2}(R^3)$ (see Ref. [3], Theorems 2, 6).

The well-known classical results belonging to O. A. Ladyzhenskaya and J. Serrin (see Refs. [1, 2]) show an existence of time interval $[0, T)$ where the solution of problem (1) is regular in zone $[0, T) \times R^3$. Denote by $T_\star$ the least upper bound of these T. If $T_\star < \infty$, then the solution of problem (1) is called a blow up solution, and the fluid flow describing of this solution is called a turbulent flow. In this case, we have the collapse phenomenon.

Now, describe the control parameters knowing the values of which we can be sure that there will be no collapse or it will not be in a guaranteed time interval. Following [3] (see formulae (68), (69), (87), and (5) from [3] respectively), we define these parameters $\lambda, \mu, \varepsilon$ as parameters, which control solution smoothness and a number T_0 by equalities:

$$l(\varphi) = \|\varphi\|_2 \cdot \|\nabla \varphi\|_2, \quad \lambda = \left(\frac{4\sqrt[4]{3}}{3a_1}\right)^2 \frac{\nu^2}{l(\varphi)} = \frac{81\nu^2}{8l(\varphi)}, \quad \lambda(t) = \frac{81\nu^2}{8l(u(t, \cdot))}, \tag{3}$$

where $u(t, x)$ is the solution of Cauchy problem (1),

$$\mu = \frac{T_\star}{T_0}, \tag{4}$$

$$\|u(T_0, \cdot)\|_2^2 = \|\varphi\|_2^2(1 - \varepsilon\lambda^2), \tag{5}$$

$$T_0 = \left(\frac{9}{4}\right)^4 \frac{\nu^3}{\|\nabla \varphi\|_2^4}. \tag{6}$$

Here, $[0, T_0)$ is that time interval (it is not necessarily optimal) where every weak solution (see definition in Refs. [1, 4, 6]) of problem (1) is regular (i.e., smooth) and it satisfies condition:

$$\|\nabla u(t, \cdot)\|_2^2 \leq \frac{\|\nabla \varphi\|_2^2}{\sqrt{1 - \frac{t}{T_0}}} \tag{7}$$

(see Lemma 39 from Ref. [3]).

If $T_\star$ is finite (see Ref. [3], Lemma 50, Theorems 6–7), then for these parameters, there are fulfilled inequalities: $\lambda < 1, 0 < \varepsilon < 1$ and

$$\tau^2(\varepsilon) = \frac{1}{4}\left(\varepsilon + \frac{1}{\varepsilon}\right)^2 < \mu < \lambda^{-4}. \tag{8}$$

From Leray's estimates (see Ref. [8]), it follows that every blow up solution of problem (1) satisfies the condition:

$$\int_0^{T_\star} \|\nabla u(t, \cdot)\|_2^4 dt = +\infty \tag{9}$$

for finite T_*. Nevertheless, this weak solution for every $T < T_*, T > 0$, satisfies the inequality:

$$\int_0^T \|\nabla u(t, \cdot)\|_2^4 dt < +\infty, \tag{10}$$

that implies solution smoothness on set $[0, T] \times R^3$ (see Ref. [2].)

3. Main results

If initial data $\varphi \in C_{6/5,3/2}^\infty (R^3)$ and parameter $\lambda \geq 1$, then Cauchy problem (1) has a global regular solution (see Theorem 7 from Ref. [3]). In other words, we have no any collapse in this case. Note, from the definition of parameter λ, condition $\lambda \geq 1$ means that the rate of change of kinetic energy square at initial point $t = 0$ is negligible. It implies the determinism from the begining. If it is not there, we introduce new parameters and tools. Therefore, the following two results are very useful to compare with formula (9). The first of them it gives the sufficient condition of determinism by another tool. Moreover, it is important to note here that we do not require finiteness of mixed norm

$$\|u\|_{p,q} = \left(\int_0^T \|u(t, \cdot)\|_p^q dt \right)^{1/q}, \quad p \geq 3, \quad q \geq 2, \quad 3/p + 2/q \leq 1. \tag{11}$$

over time interval $[0, T]$, where Lebesgue norm is defined by formula (2). Every weak solution of problem (1) with finite mixed norm is smooth and unique (see Refs. [1, 2, 4, 6]).

Theorem 1. Let $\varphi \in C_{6/5,3/2}^\infty (R^3)$ be initial data of problem (1). Parameter $\lambda < 1$ and mean T_0 are defined by formulae (3) and (6), respectively. Suppose that

$$\frac{1}{T_0} \int_0^{T_0} \|\nabla u(t, \cdot)\|_2^4 dt \leq \|\nabla \varphi\|_2^4 \ln \frac{1}{1 - \lambda^4}, \tag{12}$$

where u is a smooth solution of problem (1) on the time interval $[0, T_0)$. Then, this solution has a global regular extension on the set $[0, \infty) \times R^3$. Moreover, the following estimates are fulfilled:

$$\|\nabla u(T_0, \cdot)\|_2^2 \leq \frac{\lambda^2 \|\nabla \varphi\|_2^2}{\sqrt{1 - \lambda^4}}, \quad \lambda(T_0) > 1,$$

$$\|\nabla u(t, \cdot)\|_2^2 \leq \frac{\lambda^2(T_0)}{\lambda^2(T_0) - 1} \|\nabla u(T_0, \cdot)\|_2^2 \tag{13}$$

for all $t > T_0$, where $\lambda(T_0)$ is defined in formula (3).

Proof. For the first time, we note that there exists a number $\xi \in (0, \; T_0]$ satisfying inequality

$$\|\nabla u(\xi, \cdot)\|_2^2 \leq \frac{\lambda^2 \|\nabla \varphi\|_2^2}{\sqrt{1 - \lambda^4 \frac{\xi}{T_0}}}. \tag{14}$$

Let us suppose the opposite. Then, on interval $[0, \; T_0]$, we have the following inequality:

$$\|\nabla u(t, \cdot)\|_2^4 > \frac{\lambda^4 \|\nabla \varphi\|_2^4}{1 - \lambda^4 \frac{t}{T_0}}. \tag{15}$$

Integrating it over this interval, we obtain the estimate contradicting theorem condition for the mean value.

Therefore, Eq. (14) is true. Rewriting it by the following way:

$$\frac{T_0}{\lambda^4} \leq \xi + \frac{c\nu^3}{\|\nabla u(\xi, \cdot)\|_2^4} = \tau_1(\xi), c = \left(\frac{9}{4}\right)^4, \tag{16}$$

we note that, for all $t \geq \xi, t < T_*$, $\tau_1(\xi) \leq \tau_1(t)$ because function τ_1 is not decreasing (see Lemma 45, formula (85) from Ref. [3]).

Therefore, if T_* is finite, then we obtain inequality $\frac{1}{\lambda^4} \leq \mu$. This contradicts to inequality (8). Hence, $\mu = \infty$. Then, solution u of problem (1) is global and regular.

From formula (16) and monotonicity of function τ_1, it follows immediately that, for all $t, \xi \leq t < T_*$, the next inequality is fulfilled:

$$\|\nabla u(t, \cdot)\|_2^2 \leq \frac{\lambda^2 \|\nabla \varphi\|_2^2}{\sqrt{1 - \lambda^4 \frac{t}{T_0}}}. \tag{17}$$

Hence, we have the first inequality of Theorem 1.

Let us prove the second estimate. Suppose the opposite. Then, $\lambda(T_0) \leq 1$. In this case, for solution $u(t, x)$, function $\tau_2(t) = \|u(t, \cdot)\|_2^2 (\lambda^2(t) - 1)$ is not decreasing function (see Ref. [3], inequality (77)). Hence, for $0 \leq t \leq T_0$, we obtain estimates:

$$\|u(t, \cdot)\|_2^2 (\lambda^2(t) - 1) \leq \|u(T_0, \cdot)\|_2^2 (\lambda^2(T_0) - 1) \leq 0. \tag{18}$$

Then, $\lambda(t) \leq 1$.

Hence, for all t, $0 \leq t \leq T_0$, we have the estimate:

$$4c\nu^4 \leq \|u(t, \cdot)\|_2^2 \|\nabla u(t, \cdot)\|_2^2 \tag{19}$$

where constant c from Eq. (16). It is the strong inequality in some neighborhood of point $t = 0$ because $\lambda(0) = \lambda < 1$ and functions $\eta_1(t) = \|\nabla u(t, \cdot)\|_2$, $\eta_4(t) = \|u(t, \cdot)\|_2$ are continuous (see Lemma 36 in Ref. [3]).

Therefore, integrating Eq. (19) over interval $[0, T_0]$, we extract a strong estimate:

$$4c\nu^4 T_0 < \int_0^{T_0} \|u(t, \cdot)\|_2^2 \|\nabla u(t, \cdot)\|_2^2 dt = \frac{1}{4\nu} \left(\|\varphi\|_2^4 \| - \|u(T_0, \cdot)\|_2^4 \right). \tag{20}$$

Hence,

$$\|u(T_0, \cdot)\|_2^4 < \|\varphi\|_2^4 (1 - \lambda^4). \tag{21}$$

Apply this inequality and inequality (17) for $t = T_0$, then from (19), we obtain the strong estimate

$$4c\nu^4 < \|\varphi\|_2^2 \|\nabla\varphi\|_2^2 \lambda^2 = 4c\nu^4. \tag{22}$$

Contradiction. The second inequality is proved.

The third estimate follows from Theorem 10 (see Ref. [3]) if we consider $u(t + T_0, x)$ as the Cauchy problem solution with initial data $u(T_0, x)$. Theorem 1 is proved.

The following statement is connected with a local extension. The model is the same as above (see Theorem 1).

Theorem 2. Let $\varphi \in C^\infty_{6/5,3/2}(R^3)$ be initial data in problem (1). Parameter $\lambda < 1$ and mean T_0 are defined by formulae (3) and (6), respectively. Suppose that

$$\frac{1}{T_0} \int_0^{T_0} \|\nabla u(t, \cdot)\|_2^4 dt \leq \|\nabla\varphi\|_2^4 \ln\frac{1}{1 - \lambda^2}, \tag{23}$$

where u is a smooth solution of problem (1) on time interval $[0, T_0)$. Then, this solution has a local smooth extension on set $\left[0, \frac{T_0}{\lambda^2}\right) \times R^3$. In addition, it is true the following estimate:

$$\|\nabla u(t, \cdot)\|_2^2 \leq \frac{\lambda\|\nabla\varphi\|_2^2}{\sqrt{1 - \lambda^2 \frac{t}{T_0}}} \tag{24}$$

for every moment $t, T_0 \leq t < \frac{T_0}{\lambda^2}$.

Proof. This theorem is proved by the same way as Theorem 1. Here, there exists a number $\xi \in (0, \ T_0]$ satisfying condition

$$\|\nabla u(\xi, \cdot)\|_2^2 \leq \frac{\lambda\|\nabla\varphi\|_2^2}{\sqrt{1 - \lambda^2 \frac{\xi}{T_0}}}. \tag{25}$$

As in Theorem 1 above, we prove it from the opposite. The monotonicity of function τ_1 from proof of this theorem implies inequality:

$$\frac{T_0}{\lambda^2} \leq \xi + \frac{c\nu^3}{\|\nabla u(\xi, \cdot)\|_2^4} = \tau_1(\xi) \leq \tau_1(t), \ \ \xi < t. \tag{26}$$

Hence, we have estimate (24). Theorem 2 is proved.

Remark. The number $\frac{T_0}{\lambda^2} = \frac{\|\varphi\|_2^2}{4\nu\|\nabla\varphi\|_2^2}$ is interesting because it does not depend on from constants in apriori estimates for solutions. This is the first. Second, it influences on estimates for kinetic energy of turbulence flows at moment close to initial (see Ref. [9], Theorem 1).

Theorem 3. Let $\varphi \in C^\infty_{6/5,3/2}(R^3)$ be initial data in problem (1). Parameter $\lambda < 1$ and $\sqrt{2}\lambda^2 \geq 1$. In addition, number T_0 and parameter ε are defined by formulae (5) and (6), respectively. Suppose $\varepsilon \leq \sqrt{2} - 1$. Then, a weak solution u of problem (1) is global and regular.

Proof. If solution u of problem (1) is blow up solution, then from Eq. (8) and theorem conditions, we obtain for parameter μ following inequalities: $\mu < \lambda^{-4} \leq 2$. Hence, and still one estimate from Eq. (8), we have inequality: $\frac{1}{2}\left(\varepsilon + \frac{1}{\varepsilon}\right) < \sqrt{2}$.

Therefore, $\varepsilon > \sqrt{2} - 1$. Contradiction. Theorem 3 is proved.

Theorem 4. Let $\varphi \in C^{\infty}_{6/5,3/2}(R^3)$ be initial data in problem (1). Parameters $\lambda < 1$ and ε are defined by formulae (3) and (6), respectively. Suppose that number T_0 from (5) and a weak solution u of problem (1) satisfies inequality

$$\left\| u\left(\frac{T_0}{\lambda^4}, \cdot\right) \right\|_2^2 \geq \|\varphi\|_2^2 (1 - \tau(\varepsilon)\lambda^2), \tag{27}$$

where function $\tau(\varepsilon)$ from Eq. (8). Then, solution u is global and regular.

Proof. Suppose the opposite. Then,

and
$$T_* < \frac{T_0}{\lambda^4}$$

$$\|u(T_*, \cdot)\|_2^2 \geq \|\varphi\|_2^2 (1 - \tau(\varepsilon)\lambda^2). \tag{28}$$

On the other hand, we have

$$\|u(T_*, \cdot)\|_2^2 \leq \|\varphi\|_2^2 (1 - \sqrt{\mu}\lambda^2). \tag{29}$$

(see Ref. [3], Lemma 49). Comparing them, we obtain the contradiction with (8). Theorem 4 is proved.

Theorems 3 and 4 describe the influence on the appearance of collapse by kinetic energy estimates. For application of them, the other facts are considered lower.

4. Kinetic energy estimates and control of solution smoothness

Now, we consider the effect of kinetic energy estimates on the lifetime of smooth solution u in problem (1). Choose T_0 from Eq. (6), $t \in (0, T_0)$ and integrate (7) over $[0, t]$. Then, we obtain

$$\| u(t, \cdot) \|_2^2 \geq \|\varphi\|_2^2 \left(1 - \lambda^2 + \lambda^2 \sqrt{1 - \frac{t}{T_0}}\right) \geq \|\varphi\|_2^2 \left(1 - \lambda^2 \sqrt{\frac{t}{T_0}}\right) \tag{30}$$

(see Lemma 41 in Ref. [3]). If mean $\| u(T_0, \cdot) \|_2^2$ is not close to minimum, then we have no blow up solution. This is described by the following theorem.

Theorem 5. Let $\varphi \in C^{\infty}_{6/5,3/2}(R^3)$ be initial data in problem (1). Parameter $\lambda < 1$ and mean T_0 are defined by formulae (3) and (6), respectively. Suppose that

$$\| u(T_0, \cdot) \|_2^2 \geq \|\varphi\|_2^2 \sqrt{1 - \lambda^4}, \tag{31}$$

where u is a smooth solution of problem (1) on the time interval $[0, T_0)$. Then, solution u can be extended as the global and regular.

Proof. Suppose the opposite. Then, for every $t \in [0, T_0)$, it is fullfilled inequality $\lambda(t) < 1$ (see (3)) for finitite mean T_* (see Ref. [3], Lemma 50, Theorems 6–7). Then, we have (19)) where the inequality is strong for all t. Hence, integrating it over $[0, T_0)$, we obtain (21). This contradicts the condition from Theorem 5. It is proved.

For the first time, this statement is given in Ref. [3].

The following theorem shows lifetime of smooth solution in problem (1) if the kinetic energy dissipation is increasing.

Theorem 6. Let $\varphi \in C_{6/5,3/2}^\infty(R^3)$ be initial data in problem (1). Parameter $\lambda < 1$ and mean T_0 are defined by formulae (3) and (6), respectively. Suppose that

$$\|u(T_0, \cdot)\|_2^2 \geq \|\varphi\|_2^2 \left(1 - \lambda + \lambda\sqrt{1 - \lambda^2}\right), \tag{32}$$

where u is a smooth solution of problem (1) on the time interval $[0, T_0)$. Then, solution u can be extended a local and regular on set $\left[0, \frac{T_0}{\lambda^2}\right) \times R^3$.

Proof. We introduce a function

$$\omega(t) = \|u(t, \cdot)\|_2^2 - \|\varphi\|_2^2 \left(1 - \lambda + \lambda\sqrt{1 - \lambda^2 \frac{t}{T_0}}\right). \tag{33}$$

Since function ω is smooth and $\omega(0) = 0$, $\omega(T_0) \geq 0$, then there exists a number $\xi \in (0, T_0)$ satisfying condition $\omega'(\xi) \geq 0$. Hence, it follows inequality (25), which implies (26). From monotonicity of function τ_1 in formula (26), we get the main statement. Theorem 6 is proved.

Finally, for illustration, we show some facts due to blow up solutions and kinetic energy (other aspects connecting with estimates of norm $\|u(t, \cdot)\|_p, p \geq 3$, are covered in Refs. [4, 5, 8]).

Theorem 7. Let $\varphi \in C_{6/5, 3/2}^\infty(R^3)$ be initial data. Suppose parameter $\lambda < 1$ and u is blow up solution of problem (1). If there exists a number t_0 from interval $(0, \tau^2(\varepsilon)T_0)$ with condition

$$\|u(t_0, \cdot)\|_2^2 = \|\varphi\|_2^2 \sqrt{1 - \frac{t_0}{\mu T_0}}, \tag{34}$$

where function $\tau = \tau(\varepsilon)$ from Eq. (8). Then, parameter μ from formula (4) satisfies the estimate:

$$\mu \geq \frac{1}{\lambda^2(2 - \lambda^2)}, \tag{35}$$

that is, solution u is regular on strip domain $\left[0, \frac{T_0}{\lambda^2(2-\lambda^2)}\right) \times R^3$.

The proof is given in Ref. [9], Theorem 1.3.

Here, we must note the critical level of kinetic energy. Naturally, it is described by equality

$$\|u(t_0, \cdot)\|_2^2 = \|\varphi\|_2^2 \left(1 - \lambda^2\sqrt{\frac{t_0}{T_0}}\right) \tag{36}$$

because the last inequality in Eq. (30) and the following theorem are true (see Ref. [3], Theorem 8).

Theorem 8. Let $\varphi \in C^{\infty}_{6/5,\ 3/2}(R^3)$ be initial data. Suppose parameter $\lambda < 1$ and a vector field u is a solution of problem (1), on an interval $[0, T)$, inequality

$$\|u(t, \cdot)\|_2^2 \geq \|\varphi\|_2^2 \left(1 - \lambda^2 \sqrt{\frac{t}{T_0}}\right) \tag{37}$$

is fullfilled. Then, the weak solutions $u(t, x) = (u_1(t, x), u_2(t, x), u_3(t, x))$ and $P = P(t, x)$ are regular on strip domain $[0, T) \times R^3$.

Proof. Suppose $T_* < T$. Then, from condition, it follows

$$\|u(T_*, \cdot)\|_2^2 \geq \|\varphi\|_2^2 (1 - \sqrt{\mu} \lambda^2). \tag{38}$$

Blow up solution of problem (1) satisfies inequalities:

$$\|\varphi\|_2^2 \lambda^2 \sqrt{\mu - \frac{t}{T_0}} + \|u(T_*, \cdot)\| \leq \|u(t, \cdot)\|_2^2 \leq \|\varphi\|_2^2 \left(1 - \sqrt{\mu} \lambda^2 + \lambda^2 \sqrt{\mu - \frac{t}{T_0}}\right) \tag{39}$$

(see Ref. [3], Lemma 49). Comparing these estimates, we get identity:

$$\|u(t, \cdot)\|_2^2 = \|\varphi\|_2^2 \left(1 - \sqrt{\mu} \lambda^2 + \lambda^2 \sqrt{\mu - \frac{t}{T_0}}\right). \tag{40}$$

Differentiate this identity at point $t = 0$. After simple calculations, we obtain equality $\mu = 1$. This contradicts to the left-hand inequality in Eq. (8) because dissipation parameter $\varepsilon < 1$ (see Ref. [3], Lemma 44). The smoothness pressure function P follows from inclusion $u(t, \cdot) \in C^{\infty}_{6/5,\ 3/2}$ (see Ref. [3], Theorem 6). The theorem is proved.

5. Conclusions

Summarizing above, we underline the first controlling parameter λ defined at the initial time by formula (3). If this parameter $\lambda \geq 1$, then we have no blow up solution.

Next, if $\lambda < 1$, then we introduce another controlling parameter ε which is defined at point time T_0 by formula (5). If at this point time, the kinetic energy satisfies the inequality

$$\|u(T_0, \cdot)\|_2^2 \geq \|\varphi\|_2^2 \sqrt{1 - \lambda^4} \tag{41}$$

then we do not have phenomenon collapse again, as in the previous case. But if the kinetic energy changes in the boundaries

$$\|\varphi\|_2^2 \left(1 - \lambda + \lambda \sqrt{1 - \lambda^2}\right) \leq \|u(T_0, \cdot)\|_2^2 < \|\varphi\|_2^2 \sqrt{1 - \lambda^4} \tag{42}$$

then the guaranteed time interval without collapse is $\left[0, \frac{T_0}{\lambda^2}\right)$. In general, it is given by formula (8).

A more refined result is connected with the following lower estimate. If kinetic energy satisfies inequality

$$\|u(t, \cdot)\|_2^2 \geq \|\varphi\|_2^2 \left(1 - \lambda^2 \sqrt{\frac{t}{T_0}}\right) \tag{43}$$

for every t, $0 \leq t \leq T$, then the weak solution u of problem (1) is regular on time interval $[0, T)$. Now, no examples where kinetic energy could satisfy the strong opposite inequality:

$$\|u(t, \cdot)\|_2^2 < \|\varphi\|_2^2 \left(1 - \lambda^2 \sqrt{\frac{t}{T_0}}\right) \tag{44}$$

yet. Therefore, it may be perspective for the further researches. This is the first.

Second, the control may not be point time-based, but on average. For example, if parameter $\lambda < 1$ and for velocities gradient, the mean value satisfies the inequality

$$\frac{1}{T_0} \int_0^{T_0} \|\nabla u(t, \cdot)\|_2^4 dt \leq \|\nabla \varphi\|_2^4 \ln \frac{1}{1 - \lambda^4}, \tag{45}$$

then we have no collapse. The existence of smooth solution on time interval $\left[0, \frac{T_0}{\lambda^2}\right)$ is described by the inequality of Theorem 2.

In other words, we are doing a subsequent smmothness control in the same way as we are finding optimality control in Bellman principle. Here, the controlling parameters may be various, and, therefore, the researches will be interesting also in this way.

The most important results are described by Theorems 1–2, 5–6, and 8.

Acknowledgements

No external funding was received for this study.

Conflict of interest

The author declares no conflict of interest.

Notes/thanks/other declarations

Dedicated to the bright memory of Academician Ju. G. Reshetnyak.
The author is grateful to all reviewers for their critical comments. Many thanks.

Author details

Vladimir I. Semenov
Immanuel Kant Baltic Federal University, Kaliningrad, Russia

*Address all correspondence to: visemenov@rambler.ru

IntechOpen

References

[1] Ladyzhenskaya OA. Mathematical Questions of Dynamics of Viscous Incompressible Fluid. 2nd ed. Moscow, Russia: Nauka; 1970. (In Russian)

[2] Serrin J. On the interior regularity of weak solutions of Navier–Stokes equations. Archive for Rational Mechanics and Analysis. 1962;**9**:187-195

[3] Semenov VI. The 3d Navier-Stokes equations: Invariants, local and global solutions. Axioms. 2019;**8**(41):1-51. Available from: http://mdpi.com/journal/axioms

[4] Escauriaza L, Seregin GA, Sverak V. $L_{3, \infty}$ solutions to the Navier-Stokes equations and backward uniqieness. Uspekhi Matematicheskih Nauk. 2003; **58**:3-44. (In Russian)

[5] Seregin G. A certain necessary condition of potential blow up for Navier–Stokes equations. arXiv. 2011, arXiv:1104.3615:1-16. Available from: http://arxiv.org/pdf/1104.3615

[6] Galdi GP. An Introduction to the Mathematical Theory of the Navier–Stokes Equations, Steady Problems. 2nd ed. New York: Springer; 2011

[7] Dobrokhotov SJ, Shafarevich AI. Some integral identities and remarks on the decay at infinity of the solutions to the Navier-Stokes equations in the entire space. Russian Journal of Mathematical Physics. 1993;**2**(1):133-135

[8] Leray J. Sur le mouvement d'un liquide visqueux emplissant l'espace. Acta Math. 1934;**63**:193-248. (In French)

[9] Semenov VI. Some properties of blow up solutions in the Cauchy problem for 3d Navier-Stokes equations. Symmetry. 2020;**9**(1523):1-7. Available from: http://mdpi.com/journal/symmetry

Chapter 3

Stability of Vortex Symmetry at Flow Separation from Slender Bodies and Control by Local Gas Heating

Vladimir Shalaev

Abstract

A new approach to describe the asymmetry vortex state occurrence for the separated flow over slender bodies is presented. On the basis of the proposed model, a criterion of the asymmetry origin for conical bodies is found using catastrophe theory. Main properties of the transition to an asymmetric state are studied on the basis of the local analysis, the flow characteristics near the critical saddle point. Using the obtained criterion and the new model, numerical calculations of turbulent boundary layer are made to estimate an effectiveness of global flow structure control methods using local plasma discharge or surface heating. The qualitative confirmation of presented numerical results was done by experiments.

Keywords: slender bodies, separated flows, vortex instability model, control with the plasma discharge and surface heating, boundary layer model

1. Introduction

The symmetric vortex structure stability problem for the separated flow over slender bodies has important practical applications, and it is intensively investigated using numerical and experimental methods during a long enough period. Reviews of these investigations including theoretical, experimental and numerical investigations are presented in works [1–15]. However, reasons and mechanism of a spontaneous transition to asymmetry are not defined completely up to date. It is known that this phenomenon can be initiated by different flow heterogeneities, for example, small surface deformations [5, 6], gas blow-suction through holes [7], acoustic waves and other incoming flow disturbances.

Experiments and calculations show that even very small disturbances, such as technological surface defects, a numerical grid imperfection and rounding computer errors, can lead to the flow asymmetry. Also, outer reasons of the flow asymmetry arising exist, which are connected with the laminar-turbulent transition [8] or the vortex destruction [9], but they are not considered in this work.

There are two different points of view on asymmetry arising: the convective amplification of stationary disturbances, generated near the tip [7, 11], or an absolute disturbances instability, related with the flow velocity profile in the saddle point [4, 5, 12]. However, none of these approaches allows explaining experimental observations. Also, numerical modeling results do not allow understanding the asymmetry reason [13–16]. Previous theoretical investigations based on the model vortex-cut also do not allow revealing the transition mechanism to asymmetric state [17–19].

It is assumed in the first part of the present work that the arising asymmetry is an inner property of the symmetric flow and is related with its structure (global) instability. The problem is reduced to the stationary (critical) point analysis of the nonlinear dynamic system, which is described by symmetric flow streamline equations. Strong mathematical results were obtained only for autonomic gradient systems, which correspond to conical flows. In this case, on the basis of the catastrophe theory [20], a symmetric flow stability criterion is found, which is confirmed by experimental data [4, 21–23]. On the basis of nonlinear local equation properties near the critical point, the qualitative analysis of characteristics for the transition to asymmetric state is done.

The obtained asymmetry origin criterion allowed proposing a method for the global flow structure control based on the local gas heating by the plasma discharge [1–3, 21–23]. The new model for the description of the plasma discharge in the boundary layer was proposed, and on this basis, calculations of the separation point location for the thin round cone are presented and verified using experimental data [21–23].

2. Problem formulation

Let consider the separated stationary flow around slender-pointed body (**Figure 1a** and **b**) with affine-like transverse section form $Y = F(Z)B(X)$, where $B(0) = 0$ corresponds to the body nose and X, Y, Z are nondimensional Descartes coordinates:

$$X = \frac{X^*}{l}, \quad Y = \frac{Y^*}{\delta l}, \quad Z = \frac{Z^*}{\delta l}, \quad r = \frac{r^*}{\delta l}.$$

Here, l is the body length, and $a^* = \delta l$ is its thickness. It is assumed that the flow and the body are symmetrical with respect to the plane $Z = 0$.

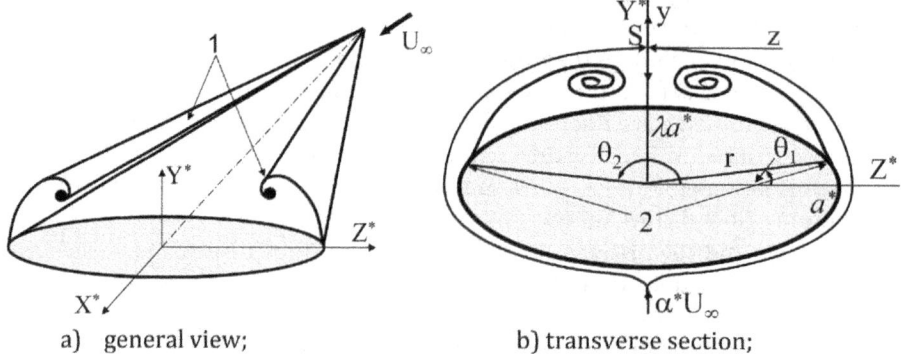

a) general view;   b) transverse section;

Figure 1.
Flow scheme. (a) General view. (b) Transverse section. 1 is the vortex sheet, and 2 is the separation location.

It is assumed that, for the Reynolds number Re, the angle of attack α^*, and the nondimensional body thickness δ, the following relations are fulfilled:

$$\delta \ll 1, \quad \delta \, \mathrm{Re} = \delta \frac{U_\infty l}{\nu_\infty} \gg 1, \quad \alpha = \frac{\alpha^*}{\delta} = O(1). \tag{1}$$

Here, U_∞ is the incoming flow speed, and ν_∞ is the kinematic viscosity.

At conditions (1), solutions of Navier-Stokes equations may be found using asymptotic expansions of flow parameters at $\delta \to 0$ and $\mathrm{Re} \to \infty$ [24]:

$$u^* = U_\infty \left[1 + \delta^2 \ln \delta u_0(X) + \delta^2 u(X, Y, Z) + \ldots \right],$$

$$w^* = U_\infty \delta w(X, Y, Z) + \ldots, \quad v^* = U_\infty \delta v(X, Y, Z) + \ldots,$$

$$p^* - p_\infty = \rho_\infty U_\infty^2 \delta^2 p(X, Y, Z) + \ldots, \quad \rho^* = \rho_\infty \left[1 + \delta^2 \rho(X, Y, Z) + \ldots \right]. \tag{2}$$

Here, u^*, v^*, and w^* are velocity components along axes X, Y, Z; p^* and ρ^* are pressure and density, and p_∞ and ρ_∞ are its non-perturbed values.

In the first-order approximation, the regular limit of Navier-Stokes equations for the asymptotic expansion (2) is the Euler equation system for the incompressible fluid in the crossflow plane with the non-flow boundary condition:

$$\frac{\partial v}{\partial X} + v \frac{\partial v}{\partial Y} + w \frac{\partial v}{\partial Z} + \frac{\partial p}{\partial Y} = 0, \quad \frac{\partial w}{\partial X} + v \frac{\partial w}{\partial Y} + w \frac{\partial w}{\partial Z} + \frac{\partial p}{\partial Z} = 0, \quad \frac{\partial v}{\partial Y} + \frac{\partial w}{\partial Z} = 0; \quad \left[v - wB \frac{dF}{dZ} \right]_{Y=BF} = F \frac{dB}{dX} \tag{3}$$

In these equations, the longitudinal coordinate X is similar to the time. Thus, stability properties of the symmetric separated flow over slender body are determined by inviscid equations. In this approximation, separation angles $\theta_1 = \theta_s(X)$ and $\theta_2 = \pi - \theta_s$ (**Figure 1b**) are problem parameters, which may be defined from boundary layer calculations or using experimental data.

For the stability analysis of the flow, we use equations for characteristics of the system (3), which are flow streamlines uniquely corresponding to each solution of Eq. (3). Streamlines are described by the following equations [25]:

$$\frac{dZ}{dX} = w(X, Y, Z), \quad \frac{dY}{dX} = v(X, Y, Z),$$

$$X = X_0 : \quad Y = Y_0, \quad Z = Z_0. \tag{4}$$

The second line of (4) specifies the initial conditions for each streamline at the plane $X = X_0$.

A stability criterion for a solution of the ordinary differential equation system (4) can be studied on the basis of the qualitative analysis of stationary points, that is, using the elementary catastrophe theory [20].

3. Stability analysis of separated conical flows

Exact mathematical results exist only for an autonomous dynamic system which for the considered case corresponds to conical flows. In this case, $B = X$, the cone surface is described by equation $\eta = F(z)$, 2δ is the maximum cone apex angle, and flow properties depend on conical variables:

$$x = \ln X, \quad \eta = \frac{Y}{X}, \quad z = \frac{Z}{X}. \tag{5}$$

The solution of Euler equation (3) can be found in the class of discontinuous potential functions, so that the following relations are valid:

$$w(\eta, z) = \frac{\partial \Phi}{\partial z}, \quad v(\eta, z) = \frac{\partial \Phi}{\partial \eta},$$

$$\frac{\partial^2 \Phi}{\partial z^2} + \frac{\partial^2 \Phi}{\partial \eta^2} = 0, \quad \left[\frac{\partial \Phi}{\partial \eta} - \frac{\partial \Phi}{\partial z}\frac{dF}{dz}\right]_{\eta=F(z)} = F.$$

Here, $\Phi(\eta, z)$ is a flow potential. The equation (5) in conical variables (6) have the following form [20]:

$$\frac{dy}{dx} = v(\eta, z) - \eta, \quad \frac{dz}{dx} = w(\eta, z) - z. \tag{6}$$

The qualitative behavior of the autonomous dynamic system (7) is determined by properties of singular critical points, where $v(X, Y, Z) = w(X, Y, Z) = 0$. The bifurcation of a solution arises at appearance of a doubly degenerated critical point, in which the Hessian of system (7) is equal to zero [20], that is,

$$\left(\frac{\partial v}{\partial \eta} - 1\right)\left(\frac{\partial w}{\partial z} - 1\right) - \frac{\partial v}{\partial z}\frac{\partial w}{\partial \eta} = 0. \tag{7}$$

Let us consider the flow scheme shown in **Figure 1b** that corresponds to separated flow at sufficiently large angles of attack. In this vortex flow, the following singular points are present: two vortex centers, the half-saddles separation points at $\theta = \theta_s$ and $\pi - \theta_s$, attachment points at $\theta = 0$ and π, and the saddle point S with coordinates $\eta = \eta_S$, $z = 0$, which is formed at the angle of attack exceeding a critical value. The first six points are Morse saddles in terms of [20], and they do not change their types at the transition to asymmetry; only the saddle point S can change its type. In the vicinity of the point S, taking into account potential and symmetry conditions, the flow velocity components are decomposed to Taylor series:

$$v = \eta_S - d\eta + b\eta^3 - 3bz^2\eta, \quad w = dz + bz^3 - 3b\eta^2 z,$$

$$y = \eta - \eta_S, \quad d(\alpha, \theta_s) = -\frac{\partial v(\eta_S, 0)}{\partial \eta} = \frac{\partial w(\eta_S, 0)}{\partial z},$$

$$b(\alpha, \theta_s) = \frac{1}{6}\frac{\partial^3 v(\eta_S, 0)}{\partial \eta^3} = \frac{1}{6}\frac{\partial^3 w(\eta_S, 0)}{\partial z^3}, \quad \frac{\partial v(\eta_S, 0)}{\partial z} = \frac{\partial w(\eta_S, 0)}{\partial \eta} = 0.$$

Near the point S, Eq. (7) are reduced to the form:

$$\frac{dy}{dx} = -cy + by^3 - 3bz^2 y, \quad \frac{dz}{dx} = -c_1 z + bz^3 - 3by^2 z,$$

$$c_1(\alpha, \theta_s) = 1 - d(\alpha, \theta_s), \quad c(\alpha, \theta_s) = 1 + d(\alpha, \theta_s). \tag{8}$$

For Eq. (9), the symmetric flow stability criterion (8) is reduced to the relation:

$$c(\alpha, \theta_s)c_1(\alpha, \theta_s) = 0. \tag{9}$$

Thus, at a bifurcation point, one of the coefficients is equal to zero: $c_1 = 0$ or $c = 0$, that corresponds to $d = 1$ or $d = -1$.

4. Stability criterion calculation and its verification by comp

To simplify calculations, the separated vortex flow over a slender cone is simulated using the model of point vortexes and conformal mapping of the body cross-section on the unit circle. In this model, the separation vortices and the vortex sheets are mimicked by the point vortices and direct line cuts connecting separation points with these vortices. The Kutta-Zhukovsky condition at the separation points [25] and the equilibrium condition for the vortex system lead to three nonlinear algebraic equations for the vortex intensity γ and vortex coordinates η_v, z_v:

$$P_i(\mathbf{x}; \theta_s, \alpha) = 0, \quad i = 1, 2, 3, \quad \mathbf{x} = (\gamma, \eta_v, z_v).$$

The separation angle θ_s and angle of attack α are parameters in this problem. Calculations were conducted for elliptic cones with the minor axis λX, $0 < \lambda \leq 1$: $\lambda = 1$ and 0 correspond to a round cone and a thin delta wing, respectively.

Typical calculations of coefficients $c(\alpha)$ (**Figures 2a and 3a**), $c_1(\alpha)$ and $b(\alpha)$ (**Figures 2b and 3b**) for a round cone ($\lambda = 1$) are presented in **Figure 2** for the laminar boundary layer separation ($\theta_s \approx 0°$) and in **Figure 3** for the turbulent boundary layer separation ($\theta_s \approx 40°$).

Parametric studies showed that coefficients $c_1(\alpha, \theta_s)$ and $b(\alpha, \theta_s)$ are always positive, and only $c(\alpha, \theta_s)$ changes its sign for each separation angle when the angle of attack is varied. From these properties, it follows that bifurcation criterion can be written as follows:

$$c(\alpha, \theta_s) = 0. \tag{10}$$

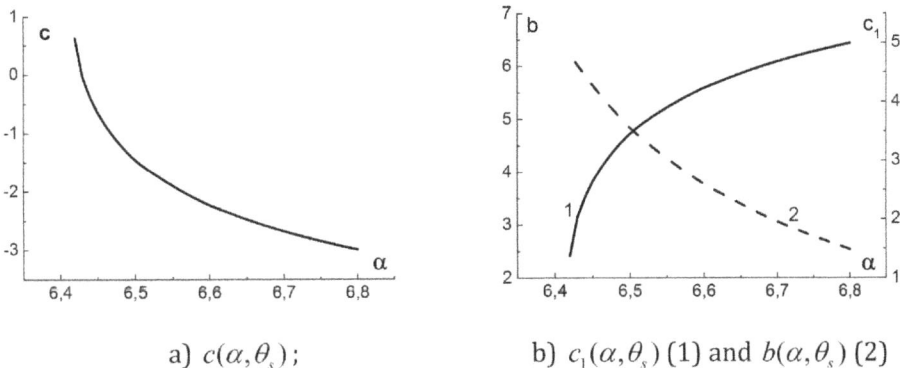

a) $c(\alpha, \theta_s)$;

b) $c_1(\alpha, \theta_s)$ (1) and $b(\alpha, \theta_s)$ (2)

Figure 2.
Coefficients of Eq. (8) for round cone ($\lambda = 1$) at laminar flow separation ($\theta_s \approx 0°$).

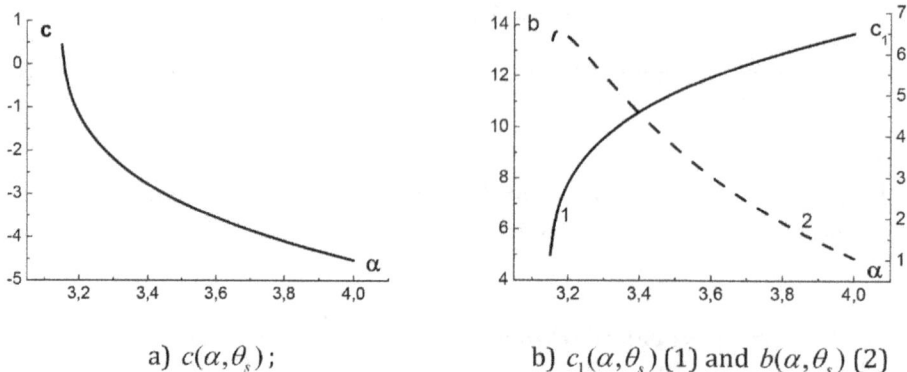

a) $c(\alpha,\theta_s)$;

b) $c_1(\alpha,\theta_s)$ (1) and $b(\alpha,\theta_s)$ (2)

Figure 3.
Coefficients of the Eq. (8) for round cone ($\lambda = 1$) at turbulent flow separation ($\theta_s \approx 40°$).

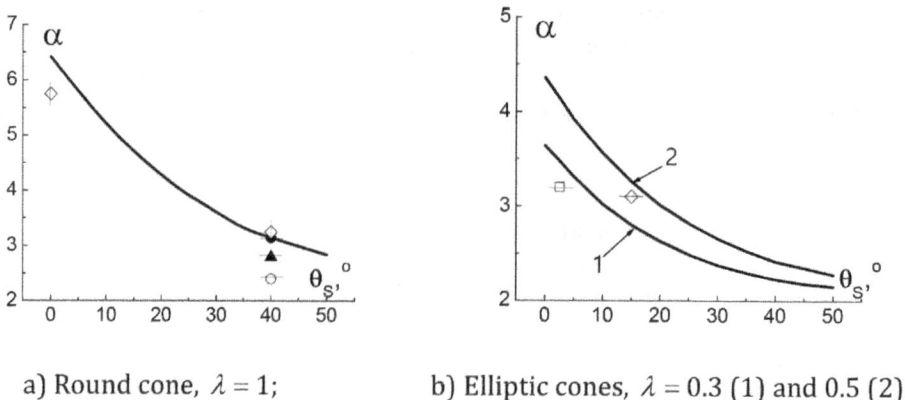

a) Round cone, $\lambda = 1$;

b) Elliptic cones, $\lambda = 0.3$ (1) and 0.5 (2)

Figure 4.
The stability criteria $c(\alpha, \theta_s) = 0$ for round (a) and elliptic (b) cones.

In **Figure 4a** and **b**, calculated curves $c(\alpha, \theta_s) = 0$ for round ($\lambda = 1$, **Figure 4a**) and elliptic (in **Figure 4b**, line 1 corresponds to $\lambda = 0.3$, and line 2 corresponds to $\lambda = 0.5$) cones are shown by solid lines. Symbols correspond to experimental data of asymmetric state origin; vertical and horizontal segments show measurement accuracy of normalized angle of attack and separation angle.

Only data for homogeneous flow conditions (laminar or turbulent) are used for the calculation verification, a transition regime is excluded according to the state diagram similar to the diagram [9] for the body of the type revived cylinder. For elliptic cones, experimental data with the flow asymmetry due to the vortex explosion [10] were excluded. Data [4] for the turbulent separation on the round cone with $\delta = 5^0$ agree with data [5] obtained at the large Reynolds number ($13.5 \cdot 10^6$) and with other data obtained in low noise wind TunnelSat large Reynolds numbers [8, 9, 21–23]. Experiments [21–23] were performed in the low noise wind tunnel of ITAM SB RAS at high level of the cone surface smoothness; the turbulent flow was generated by turbulators.

The utilized asymptotic approach is appropriate at the limit $\delta \to 0$. The data for round cones of different apex angles in **Figure 4a** ($\bigcirc - \delta = 10^{\circ}$, $\blacktriangle - \delta = 5^{\circ}$, $\bullet - \delta = 3^{\circ}$) [4] demonstrate the well convergence of the theory and experiments with the cone apex angle diminishing for the turbulent boundary layer separation ($\theta_s \approx 40°$). For the thinner cone ($\delta = 3°$), experimental results are absent; in this case, the extrapolation of [4] is used, and the obtained value is agreed well with calculation results. Calculations conform well by experiments [21–23] (**Figure 4a**, $\Diamond - \delta = 5°$), as for turbulent ($\theta_s \approx 40°$) and the laminar ($\theta_s \approx 0°$) separations. In spite of the rough enough flow model and large apex angle ($\delta = 10°$), the calculated stability criteria in **Figure 4b** for elliptic cones with $\lambda = 0.3$ (1) and $\lambda = 0.5$ (2) are consistent satisfactorily with experiments [4].

It should be noted that, in contrast to presented results, the bifurcation diagram of the intrinsic asymmetric solutions of the model [17] does not correspond the experimental data, the saddle point appears in the flow at significantly large angles [1–4, 21–23].

5. Transition to asymmetry local analysis

Although the catastrophe theory [20] gives only the criterion of the asymmetry origin but cannot describe the transition process, some qualitative results of the beginning of this process can be obtained by the analysis of local streamline Eq. (8). The transition to asymmetric state is associated with the typical change of the critical point S, when the coefficient $c(\alpha, \theta_s)$ changes its sign, and with the global flow insta-bility at $c(\alpha, \theta_s) < 0$. Calculations for elliptic cones show that $b > 0$, $c_1 > 3c$. Thus, critical points of Eq. (8) are as follows: the symmetric saddle point $y = z = 0$ (S), and also points S_1, S_2 at $c < 0$ and S_3, S_4 at $c > 0$:

$$c < 0. \quad S_1: \ y = 0, \quad z = \sqrt{\frac{c_1}{b}} \ ; \quad S_2: y = 0, \quad z = -\sqrt{\frac{c_1}{b}} \ ,$$

$$c > 0. \quad S_3: z = 0, \quad y = \sqrt{\frac{c}{b}} \ ; \quad S_4: z = 0, \quad y = -\sqrt{\frac{c}{b}} \ .$$

At $c < 0$, three nondegenerate critical points (S, S_1, and S_2) exist at the axis $y = 0$; S is unstable saddle. At $c = 0$, critical points S, S_3, and S_4 coincide, that is, S is triply degenerated in the y-direction and has the saddle-node type. At $c > 0$, there are five nondegenerate critical points: S; S_1 and S_2 at the axis $y = 0$, and S_3 and S_4 at the axis $z = 0$; in this case, S is a stable node.

For qualitative analysis of transition to asymmetry, let us consider the first equa-tion of (6) in the plane $z = 0$, where it is reduced to the simplest form of the Ginzburg-Landau equation [25]:

$$\frac{dy}{dx} = -cy + by^3. \tag{11}$$

At $c \neq 0$ and $b > 0$, Eq. (11) has the following solution:

$$h(X) = y^2 = h_c \left[1 - B_0 \left(\frac{X}{X_0} \right)^{2c_1} \right]^{-1}, \quad B_0 = \frac{h_0 - h_c}{h_0}, \quad h_c = \frac{c}{b}. \tag{12}$$

The condition $h(X_0) = h_0$ specifies an initial streamline perturbation in the plane $z = 0$.

At $c > 0$, the solution of Eq. (12) is stable in the linear approximation ($b = 0$):

$$y = y_0 \left(\frac{X}{X_0}\right)^{-c} \to 0, \quad X \to \infty.$$

For $h_0 < h_c$, the coefficient $B_0 < 0$ and the solution (12) of the nonlinear Eq. (11) is also stable: $h(X) \to 0$ at $X \to \infty$, that is, streamlines remain in the attraction region of the stable node S.

For $h_0 > h_c$, the coefficient $B_0 > 0$ and $h(X) \to \infty$ at $X \to X_c$, where

$$X_c = X_0 B_0^{-\frac{1}{2c}}, \quad Y_c = \pm X_0 \sqrt{\frac{c}{b}} = \pm X_0 \sqrt{h_c}.$$

Here, X_c is length of perturbation development; it is a linear function of the distance from the initial data plane to the top of the body X_0 and nonlinear function of the initial perturbation amplitude h_0. Thus, at $h_0 > h_c$ and $c > 0$, streamlines leave the attraction region of the stable node S at the finite distance $X = X_c$ from the cone apex. The value Y_c characterizes the threshold level of stationary perturbation at the plane $X = X_0$ that initiates instability. Y_c and X_c are decreased when the initial data plane is shifted to the cone apex: $Y_c \to 0$ and $X_c \to 0$ at $X_0 \to 0$. Therefore, at $h_0 > h_c$ and $c > 0$, the solution (12) describes the subcritical bifurcation under the influence of finite perturbations.

If $c < 0$, the solution (12) is unstable in linear and nonlinear approximations since the coefficient $B_0 > 0$ and $h(X) \to \infty$ at $X \to X_c$.

For $c = 0$, the solution of Eq. (12) has the following form:

$$h = h_0 \left[1 - 2bh_0 \ln\left(\frac{X}{X_0}\right)\right]^{-1}, \quad X_c = X_0 \exp\left(\frac{1}{2bh_0}\right)$$

This is also nonlinear unstable solution, in which the transition region depends on the initial perturbation exponentially.

The presented results show that the asymmetry origin mechanism is essentially nonlinear. At angles of attack less the critical value, the subcritical bifurcation is possible if perturbations are larger than threshold value, which is lower than closer to the top it arises. Due to the nonlinearity, it is impossible to extract perturbations from general equations and study them separately. There is a finite length where perturbations are amplified to infinity, and this distance is diminished if the initial data plane is shifted to the body apex. The subcritical bifurcation is possible at angles of attack lower than a critical value, if the initial perturbation exceeds a threshold, which decreases with shifting of the initial data plane to the body apex.

The mechanism of asymmetric flow arising has properties both convective and absolute instability, where the coordinate X is treated as a time. However, this terminology relates to the linear theory of hydrodynamic stability and cannot fully describe the considered nonlinear process. This instability can be characterized more exactly as global or structural instability. The presented qualitative analysis and the asymmetry origin criterion allow to explain most of the effects found in experiments and numerical simulations. The developed approach allows also to analyze unsteady perturbations.

6. Asymmetry control using plasma discharge and local surface heating

For analysis of these problems, calculations of 3D turbulent boundary layer equations in the local self-similar approximation on a round slender cone with the half angle at the top δ at the angle of attack $\alpha^* = \alpha\delta$ are utilized. The plasma discharge effect is modeled by the volume heat source in the boundary layer. The effect of the gas ionization is neglected since the ionization coefficient is small, of the order of 10^{-5}. The flow scheme and the coordinate system are shown in **Figure 5**. As it is seen in **Figure 5b**, the heat release acts on the boundary layer as an obstacle.

Boundary layer equations are simplified using the Dorodnitsyn-Stewardson transformation, longitudinal $f(x, \eta, \varphi)$ and transverse $g(x, \eta, \varphi)$ stream functions. Transformations have a form:

$$d\eta = \frac{\rho}{\sqrt{x}}dy, \quad y = \sqrt{\mathrm{Re}}\frac{y^*}{l}, \quad \mathrm{Re} = \frac{\rho_\infty u_\infty l}{\mu_\infty}, \quad \rho = \frac{\rho^*}{\rho_\infty},$$

$$V = -\left(\frac{3}{2}f + \frac{\partial w_e}{\partial \phi}g + w_e\frac{\partial g}{\partial \phi}\right), \quad U = \frac{\partial f}{\partial \eta} = \frac{u^*}{U_\infty}, \quad W = \frac{\partial g}{\partial \eta} = \frac{w^*}{\delta U_\infty w_e} \quad h = \frac{h^*}{h_\infty h_e}.$$

Here, $U(x, \eta, \varphi)$ and $W(x, \eta, \varphi)$ are nondimensional longitudinal (along the cone surface) and circumferential velocities, $\rho(x, \eta, \varphi)$ and $h(x, \eta, \varphi)$ are nondimensional density and enthalpy related to their values on the outer boundary, $V(x, \eta, \varphi)$ is transformed normal to the cone surface velocity, $\varphi = \theta + 0.5\pi$ is angle of cylindrical coordinates, indexes ∞ and e correspond to parameters in the freestream flow and on the outer boundary. The circular inviscid velocity is approximated by the formula $w_e = 2\alpha \sin\varphi$.

Self-similar boundary-layer equations and boundary conditions are as follows:

$$V\frac{\partial U}{\partial \eta} + w_e W\frac{\partial U}{\partial \varphi} = \frac{\partial}{\partial \eta}\left(m\frac{\partial U}{\partial \eta}\right),$$

$$V\frac{\partial W}{\partial \eta} + w_e W\frac{\partial W}{\partial \varphi} + UW + \frac{\partial w_e}{\partial \varphi}W^2 - h\left(1 + \frac{\partial w_e}{\partial \varphi}\right) = \frac{\partial}{\partial \eta}\left(m\frac{\partial W}{\partial \eta}\right),$$

$$V\frac{\partial h}{\partial \eta} + w_e W\frac{\partial h}{\partial \varphi} - Q(x, y, \varphi) = \frac{\partial}{\partial \eta}k\frac{\partial h}{\partial \eta}, \quad \rho h = 1,$$

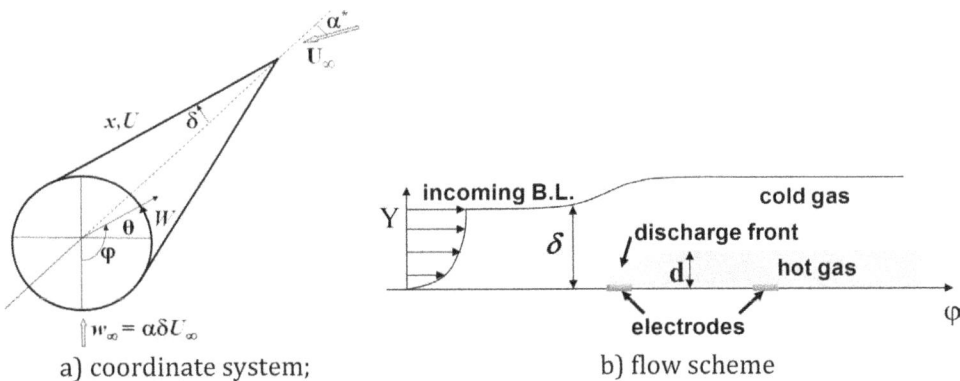

Figure 5.
Plasma discharge in the boundary layer on a round cone.

$$\eta = \infty: \quad U = W = h = 1, \quad \eta = 0: \quad U = W = 0, \quad h = h_w \left(\frac{\partial h}{\partial \eta} = 0 \right). \quad (13)$$

For turbulent flows, the transformed heat conductivity k and the transformed viscosity m are determined using the Cebeci-Smith model:

$$m = \rho\mu(1+\varepsilon), \quad k = \frac{\rho\mu}{\mathrm{Pr}} \left(1 + \frac{\mathrm{Pr}}{\mathrm{Pr}_t} \varepsilon \right); \quad 0 \le y \le y_w: \quad \varepsilon = \varepsilon_1 \left(\frac{\partial U}{\partial \eta} \right)^2,$$

$$y > y_w: \varepsilon = 0.0168\sqrt{\mathrm{Re}\,x} D_1 \frac{\delta_u}{h\mu}; \quad \varepsilon_1 = 0.16 D_1 \frac{\sqrt{\mathrm{Re}\,x}}{\mu} \left(\frac{eD}{h} \right)^2, \quad e = \int_0^{y_e} h dy,$$

$$\delta_u = \int_0^{y_e} (1-U) dy, \quad D = 1 - \exp(-d), \quad d = (\mathrm{Re}\,x)^{\frac{1}{4}} \frac{e}{26 h_w} \sqrt{\frac{1}{\mu_w} \frac{\partial U}{\partial y}};$$

$$D_1 = \left[1 + \left(\frac{e}{e_e} \right)^6 \right], \quad \mu = h^{\frac{3/2 + h_S}{h + h_S}}, \quad h_S = \frac{110.4}{T_\infty}.$$

Here, $\mathrm{Pr}_t \approx 0,9$

is the turbulent Prandtl number, and the subscript w corresponds to parameters on the wall. The heat source term Q in the left-hand side of the energy Eq. (13) describing the discharge heat release is modeled as follows:

$$Q = \frac{Q^* x l}{h_\infty U_\infty} = Q_0 y^2 \exp\left[-\frac{(y - y_c(\varphi))^2}{\sigma} \right], \quad y_c = 2 y_0 \sqrt{|(\varphi - \varphi_1)(\varphi_2 - \varphi)|}.$$

Here, Q^* and Q_0 are a dimensional source intensity and its maximum; σ characterizes the discharge width, $y_c(\varphi)$ is a centerline of the discharge that is approximated by the parabola; y_0 is a maximum distance from the discharge centerline to the wall; and the angles φ_1 and φ_2 determine electrode locations. The total power released in the boundary layer is determined as follows:

$$P = \iiint_{V_h} \rho^* Q^* a^*(x) dx^* d\phi dy^* = \frac{\beta \rho_\infty h_\infty U_\infty l^2}{\sqrt{\mathrm{Re}}} \int_{x_1}^{x_2} \sqrt{x} dx \int_{\varphi_1}^{\varphi_2} d\varphi \int_0^\infty Q dy. \quad (14)$$

If Q_0 is a function of the coordinate x and the flow is turbulent, then the boundary layer is not self-similar. However, we can use Eq. (13) for the first-order estimations of flow characteristics. Some calculation results are presented in **Figure 6** for the following parameters: $\alpha = 3.15$, $\delta = 5°$, $l = 1$ m, $T_\infty = 288.15°K$, $U_\infty = 10$ m/s, $\sigma = 1$, $y_0 = 1$; the center between electrodes is located at $\phi_0 = .5(\phi_1 + \phi_2) = 1.714$ rad (98.25°); $\phi_1 = \phi_0 - 3\Delta\phi$, $\phi_2 = \phi_0 + 3\Delta\phi$, where $\Delta\phi = 0.0314159$ is the integration step of the finite-difference approximation.

Figure 7a demonstrates the plasma discharge effect on the separation point. As the heat-source intensity increases from 0 to 400, the separation angle, φ_s, decreases from 133° to about 105°. It is seen that the plasma heating is more effective in the range $Q_0 < 100$, where the slope $d\varphi_s/dQ_0$ is relatively large.

a) temperature profiles;

b) circumferential velocity profiles;

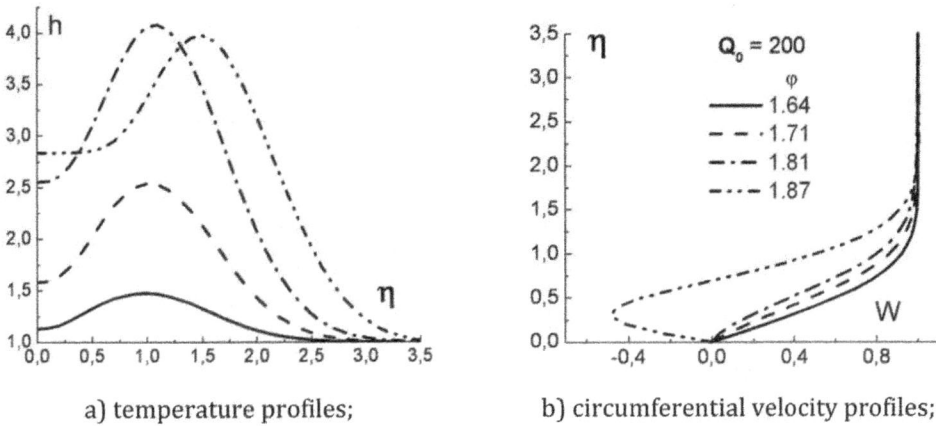

Figure 6.
Profiles of the temperature and the circumferential velocity across the boundary layer in the discharge region.

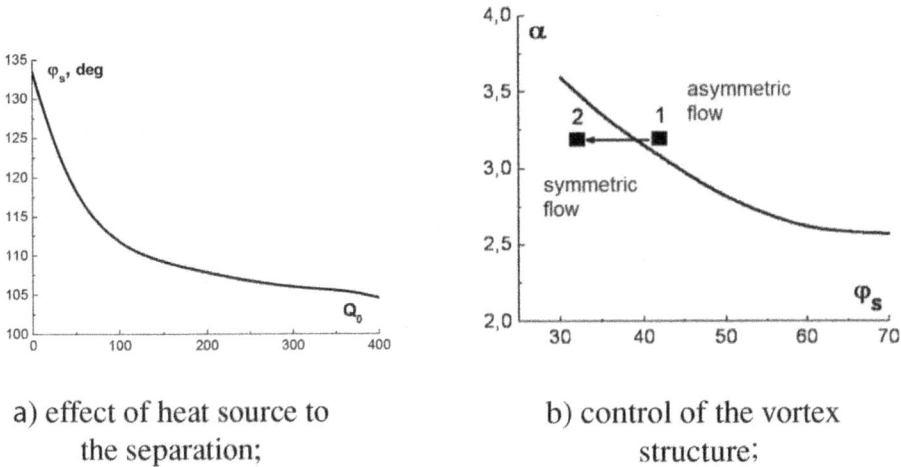

a) effect of heat source to
the separation;

b) control of the vortex
structure;

Figure 7.
Discharge effect on the separation angle and flow state.

Figure 7b illustrates feasibility of the vortex structure control using a local volume boundary-layer heating. The solid line in **Figure 4** represents boundary between symmetric and asymmetric flow states. Due to the heat source, the flow configuration changes from the initial asymmetric state ($\varphi_s \approx 133°$, symbol 1) to the symmetric state with $\varphi_s \approx 120°$ (symbol 2). This requires a nondimensional heat source intensity $Q_0 \approx 30$. Using Eq. (14), we estimate that the corresponding total power is approximately equal to 480 W. This example indicates that the method is feasible for practical applications of the global flow structure control [21–23].

Another method to control the flow separation based on the body surface local heating is also analyzed for the case of a slender round cone. The boundary layer characteristics were calculated using Eq. (13) with $Q = 0$ and with the boundary condition for enthalpy $h(0, \varphi) = h_w$, where h_w is a normalized wall temperature. The flow scheme for this case is shown in **Figure 8**. The whole cone surface had the temperature $T_w = T_\infty$ ($h_w = 1$) except the narrow strip $\Delta\varphi$ with the temperature

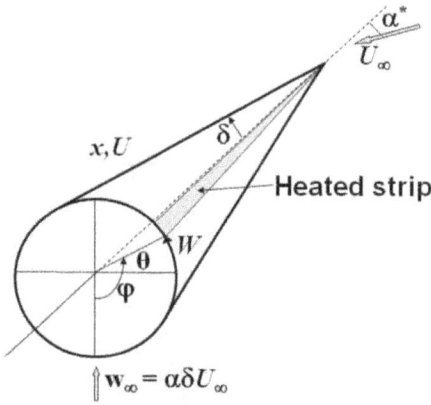

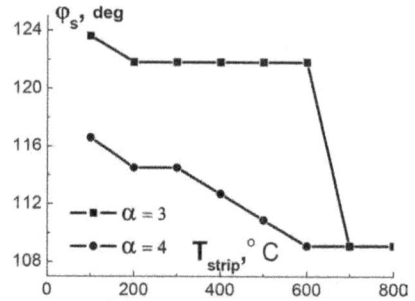

a) flow scheme;   b) effect of the strip;

Figure 8.
Effect of heated surface strip on the separation.

$T_w = T_{strip} > T_\infty$. The boundary layer calculations were conducted for the cone of length $l = 1$ m with apex angle $2\delta = 10°$ at the freestream velocity 20 m/s and normalized angles of attack $\alpha = 2 - 4$. The freestream temperature was $T_\infty = 0°C$, the strip temperature varied in the range of $T_n = 100 - 800°C$, the strip width was $\Delta\varphi = 12°$, and its center was located at $\varphi = 110°$.

The separation angles φ_s as functions of strip temperature T_{strip} are presented in **Figure 8b** for $\alpha = 3$ and $\alpha = 4$. For $\alpha = 2$, the separation angle does not depend on strip temperature. At $\alpha = 3$, the separation angle φ_s decreases from 133° to 122° with strip temperature increasing to 200°C. The separation angle is constant up to $T_{strip} = 600°C$, and then it varies to $\varphi_{1s} \approx 109°$ as strip temperature increases to 700°C and stabilizes at this value at the subsequent strip temperature growth. For $\alpha = 4$, the separation angle decreases monotonically to $\varphi_{1s} \approx 109°$, if the strip is heated from 100°C to 600°C and remains constant at further temperature growth.

The results presented in this section show that the local surface heating can be used for control of separation location; however, this method is less effective than the plasma discharge.

7. Conclusions

In this paper, two main problems related with the asymmetry arising at separated vortex flow over thin bodies are considered. In the first one, the new approach for the transition analysis to the asymmetric state appearance is proposed, and its verification on the base of the comparison of calculations with experimental results is done. In the other part, on the basis of the turbulent boundary layer, calculations and the new model for the plasma discharge possibilities of the asymmetry control using volume gas or local surface heating were considered.

The criterion of the transition to the asymmetry was obtained by applying the catastrophe theory to the analysis of the saddle critical point of the dynamical system describing symmetric separated flow streamlines. The qualitative analysis of the transition to asymmetric state properties based on the symmetric flow streamline behavior under perturbations studies near the saddle point is presented. The analysis shows that the transition is a nonlinear process, and it is characterized by the finite transition length and the threshold level of perturbations, which are diminished if the initial data plane is shifted to the cone apex. Theoretical results are verified by comparison with experimental data for laminar and turbulent flows over round and elliptic pointed cones. As shown, these results help to explain most of experimental observations and numerical simulations.

A new theoretical plasma discharge model as a local gas heat source was presented. The performed studies of separation angle control in a turbulent boundary layer using the local volume and surface gas heating show that both methods can be used for flow global structure control, but the first one is more effective.

Author details

Vladimir Shalaev
Moscow Institute of Physics and Technology (Scientific Research University),
Dolgoprudny, Russia

*Address all correspondence to: vi.shalaev@yandex.ru

IntechOpen

References

[1] Shalaev V, Fedorov A, Malmuth N, Zharov V, Shalaev I. Plasma control of forebody nose symmetry breaking. AIAA Paper. 2003:0034

[2] Shalaev V, Fedorov A, Malmuth N, Shalaev I. Mechanism of forebody nose vortex symmetry breaking relevant to plasma flow control. AIAA Paper. 2004: 0842

[3] Shalaev VI, Shalaev IV. A stability of symmetric vortex flow over slender bodies and control possibility by local gas heating. In: Reijasse P, Knight D, Ivanov M, Lipatov I, editors. EUCASS Book Series. Progress in Flight Physics. Vol. 5. Paris: EDP Sciences; 2013. pp. 155-168. ISBN:978-2-7598-0877-9. DOI: 10.1051/eucass/201305155

[4] Lowson MV, Ponton AJC. Symmetric breaking in vortex flows on conical bodies. AIAA Journal. 1992;**30**(6): 1576-1583

[5] Peake DJ, Owen FK, Johnson DA. Control of forebody vortex orientation to alleviate side forces. AIAA Paper. 1980: 0183

[6] Moskovitz CA, Hall RM, DeJarnette FR. Effects on of nose bluntness, roughness and surface perturbations on the asymmetric flow past slender bodies at large angles of attack. AIAA Paper. 1989:2236

[7] Bernhard JE, Williams DR. Proportional control of asymmetric forebody vortices. AIAA Journal. 1998; **36**(11):2087-2093

[8] Dexter PC. A study of asymmetric flow over slender bodies at high angles of attack in low turbulence environment. AIAA Paper. 1984:0505

[9] Lamont PJ. Pressure around an inclined ogive cylinder with laminar, transitional, or turbulent separation. AIAA Journal. 1982;**20**(11):1492-1499

[10] Stahl WH, Mahmood M, Asghar A. Experimental investigations of the vortex flow on delta wings at high incidence. AIAA Journal. 1992;**30**(4): 1027-1038

[11] Degani D, Tobak M. Numerical, experimental, and theoretical study of convective instability of flows over pointed bodies at incidence. AIAA Paper. 1991:0291

[12] Deng XY, Tian W, Ma BF, Wang YK. Recent progress on the study of asymmetric vortex flow over slender bodies. Acta Mechanica Sinica. 2008;**24**: 475-487. DOI: 10.1007/s10409-008-0197-3

[13] Inaba R, NishidaI H, Nonomura T, Asada K, Fujii K. Numerical investigation of asymmetric separation vortices over slender body by RANS/LES hybrid simulation. Transactions of the JSASS Aerospace Technology Japan. 2012;**10**(28):Pe_89-Pe_96

[14] Karn PK, Kumar P, Das S. Asymmetrical vortex over slender body: A computational approach. Defence Science Journal. 2021;**71**(2):282-288. DOI: 10.14429/dsj.71.15959

[15] Shahriar A, Kumar R, Shoele K. Vortex dynamics of axisymmetric cones at high angles of attack. Journal of Theoretical and Computational Fluid Dynamics. 2023;**37**:337-356. DOI: 10.1007/s00162-023-00647-0

[16] Nishioka M, Sato H. Mechanism of determination of the shedding frequency

of vortices behind a cylinder at low Reynolds numbers. Journal of Fluid Mechanics. 1978;**89**(1):49-60

[17] Dyer DE, Fiddes SP, Smith JHB. Asymmetric vortex formation from cone at incidence—A simple inviscid model. Aeronautical Quarterly. 1982;**33**(6): 293-312

[18] Kraiko AN, Reent KC. Non viscous asymmetry nature of separated flows around symmetric bodies in uniform stream. Journal of Applied Mechanics and Mathematics. 1999;**63**(1):63-70

[19] Cai J, Liu F, Luo S. Stability of symmetric vortices in two dimensions and over three-dimensional slender conical bodies. Journal of Fluid Mechanics. 2003;**480**:65-94

[20] Gilmore R. Catastrophe Theory for Scientists and Engineers. NY: Wiley Interscience Publication; 1981

[21] Fomin VM, Maslov AA, Sidorenko AA, Zanin BY, Malmuth N, et al. Control of vortex flow over bodies of revolution by electric discharge. Reports RAS. 2004;**396**(5):1-4

[22] Maslov A, Zanin B, Sidorenko A, Malmuth N, et al. Plasma control of separated flow asymmetry on a cone at high angle of attack. AIAA Paper. 2004: 0843

[23] Maslov AA, Sidorenko AA, Budovsky AD, Zanin BJ, Kozlov VV, Postnikov VV, et al. Vortex flow over conecontrol using electro spark discharge. Journal of Applied Mechanics and Technical Physics. 2010; **51**(2):81-89

[24] Cole JD. Perturbation Methods in Applied Mathematics. London: Blalsdel Publishing Comp; 1968

[25] Landau LD, Lifshits EM. Theoretical Physics. Hydrodynamics. Vol. VI. Moscow: Nauka; 1986

Section 2

Experimental and Numerical Approaches

Chapter 4

Acoustic-Vortex Decomposition Method for CFD-CAA Study of Blade Machine Noise

Sergey Timushev, Andrey Aksenov and Jiawen Li

Abstract

This chapter presents a method for calculating the generation of pressure pulsations and noise emission by blade machines based on the decomposition of the compressible medium velocity field into vortex and acoustic modes. This method of acoustic-vortex decomposition of the basic equations of motion of a compressible medium leads to an inhomogeneous wave equation with respect to enthalpy pulsations, which includes pseudo-sonic oscillations in the source vortex region and acoustic oscillations in the near-field and far-field. The source function in the wave equation is determined from the independent solution of the vortex mode equations. The boundary conditions for the wave equation are formulated using complex specific acoustic impedance and pseudo-sonic oscillations. This method allows for the consideration of the influence of inhomogeneity and turbulence of the flow, rotor interference, sound diffraction on the elements of the flow part, and impedance characteristics of the machine surfaces, while ensuring the accuracy and speed of calculations. The acoustic-vortex method represents the noise source as a function of the vortex mode velocity field. This approach eliminates the arbitrariness and conventionality of the aeroacoustic analogy, defining the source, pressure pulsations, and noise propagation in the near-field as a direct result of numerical modelling.

Keywords: acoustic-vortex decomposition, pressure pulsations, noise, BPF, CFD-CAA

1. Introduction

The reduction of noise levels and transient loads acting on the design of blade machines operating in open spaces, residential and industrial buildings, cultural institutions, car interiors, trains, aircraft and spacecraft cabins is becoming an increasingly important task. This problem is receiving increasing attention in developed countries, which is reflected in a significant tightening of noise control requirements in accordance with ISO standards [1, 2].

The application of numerical modelling and optimisation techniques to a diverse range of blade machines with subsonic flow of the working medium represents a crucial step in addressing this problem. Such machines encompass cooling fans in

IntechOpen

computer systems [3], trains [4], and modern air conditioning and air purification systems. Similar challenges arise in pumps, particularly centrifugal pumps, which represent a significant source of noise in hydraulic systems. The first references to this problem can be traced back to the 1960s when it was observed that the destruction of large industrial pumps could result in acoustic resonance [5, 6]. Similarly, the operation of small unmanned aerial systems, commonly referred to as multicopters or drones, has been found to produce noise that disturbs the public [7–10]. Furthermore, the noise generated by propellers has been linked to adverse health effects on individuals residing or working in close proximity to airports [11, 12].

The exponential growth of the UAV market for civil and military applications demonstrates the pivotal role of noise levels and propulsion efficiency in the design of modern aircraft, with these factors frequently determining the success or failure of a project [13]. Conversely, the acoustic performance of commercial propeller aircraft is becoming a crucial design parameter due to the advancement of aeromobility and distributed propulsion systems for small aircraft [14].

The primary sources of noise in subsonic flow blade machines are electromechanical or hydrodynamic in nature. Experimental studies have demonstrated [15–18] that hydrodynamic sources are the primary contributors to noise emission [19]. Furthermore, experimental experience has revealed a direct correlation between pressure pulsations and vibration load, noise emission [20, 21].

The noise of a drone is dominated by the noise of the propeller used as part of its propulsion system. The acoustic signal spectrum consists of strong harmonics manifested at frequencies that are harmonics of the blade passing frequency (BPF), as well as a broadband component [22, 23].

Noise and vibration of a hydrodynamic nature are caused by the peculiarities of the working fluid flow in the flowing part of blade machines [24–26]. These peculiarities include the unsteady interaction of the flow leaving the impeller with the stator part, which leads to the generation of pressure pulsations and noise at the blade passing frequency (BPF) and its harmonics; formation of vortices, including small-scale turbulence and large-scale vortex structures (backflow). Cavitation processes in the flowing part of pumps represent a separate source of noise [27–30], which is not considered here.

The spectral composition of pressure pulsations, noise, and vibration in a blade machine is represented by a broadband background and pronounced discrete components of the BPF. These can be amplified if they coincide with the natural resonance frequencies of the structural elements and cavities [31] or if the combination of rotor and stator blades is not optimal [32].

2. Development of computational methods

The development of computational fluid dynamics (CFD) methods has led to significant improvements in the calculation of pressure pulsations in blade machines. In particular, Croba et al. [33, 34] proposed a method for calculating unsteady two-dimensional flow in a centrifugal impeller and volute pump outlet based on the solution of averaged Navier-Stokes equations and the k-ε turbulence model.

An alternative approach utilising Reynolds equations is presented in Chu et al. [35]. The unsteady pressure in the volute outlet is derived by integrating the Reynolds equation, where all terms dependent on the velocity field are determined experimentally *via* the laser-anemometric method.

A comparable methodology is employed in the work of Thompson et al. [36], where laser-anemometry data are also utilised. However, the pressure pulsations are calculated by solving the Blokhintsev-Howe type equation [37] with respect to the stagnation enthalpy. The Lighthill equation [38] is the mainstay of current approaches to modelling blade machine noise. Subsequent works by Curl [39], Flokes-Williams, and Hawkins [40] established the theoretical basis for developing methods for calculating the aerodynamic noise of blade machines based on the aeroacoustic analogy, as well as applying Kirgoff's theorem [41]. The aeroacoustic analogy allows us to simplify the physical processes of noise generation in order to describe them analytically. At subsonic flow velocities, noise emission is mainly of the dipole source type [42]. The dipole character of radiation is caused by pressure forces acting on the rotor and stator blades from the medium flow. These forces, whether stationary or unsteady, are the cause of the BPF tone noise and its higher harmonics. They are generated by the rotor blades and, as a result of rotor-stator interaction, by the stator guide vanes. The analytical formulation of these processes is based on a formalisation proposed by Sears [43], which represents the radiated sound in the form of spiral modes [44].

Despite the findings of Tyler and Sofrin [45] that spiral modes exponentially decay at low values of Mach number of relative flow, this does not preclude their propagation at high wave numbers of flow inhomogeneity in the source.

Additional analysis is required for the case of rotor radiation in open space, where an important additional source of BPF noise generation is stationary flow inhomogeneity at the rotor inlet.

In conjunction with the advancement of computational fluid dynamics and acoustic methods, approaches based on numerical modelling of unsteady flow in bladed machines by modern computational fluid dynamics methods, such as LES with subsequent determination of acoustic radiation, are becoming increasingly prevalent [46–48]. In conjunction with the aeroacoustic analogy, other methodologies, such as RANS+ LEE + SNGR [49] and DDES [50], in conjunction with the Lighthill or Ribner equation [51], are being developed. Currently, computational fluid dynamics and acoustics methods have been extensively developed to determine the acoustic radiation of fans and propellers.

3. The source of pressure pulsations and BPF noise

The physical mechanisms of aerodynamic noise generation, particularly of axial fans, are discussed in detail in the book by Guédel [52]. He identified that fan noise consists of tonal components and broadband noise. Tonal components at blade passing frequency harmonics usually dominate the spectrum and determine the total sound power level of all types of blade machines considered.

The unsteady hydrodynamic phenomena occurring in a turbomachine can be divided into two main types (excluding cavitation). The first of these is caused by the hydrodynamic interaction of the flow leaving the working blade cascade with the stator path, which may be referred to as the "rotor-stator" interaction. The second is caused by the vortex nature of the fluid flow, which is represented in Ref. [53] the occurrence of small-scale turbulence in the boundary layer and the formation of turbulent traces when flowing around the stator elements, as well as the formation of large-scale vortex structures, back-flows under non-optimal operating modes or in an uneven velocity field at the inlet or outlet of the blade machine [54–56] are all examples of unsteady hydrodynamic phenomena.

A detailed analysis of the flow parameters in centrifugal fans [57, 58], compressors [59–61] and the flow in absolute and relative motion at the impeller outlet of centrifugal pumps [62, 63] has revealed that the flow in the blade channel and at the impeller outlet can be divided into two distinct regions: a high-energy jet and a low-energy wake zone [64, 65]. This character of the flow results in a significant non-uniformity of the relative and absolute velocities and angles of the flow along the pitch of the impeller, due to the proximity of the low-energy zone to the suction side of the blade.

The phenomenology of the generation of BPF oscillations can be analysed using a simple model shown in **Figure 1**.

Consider a semi-infinite pipe of transverse dimension L, at the inlet of which there is a "comb" of non-uniform velocity profile C with period l much smaller than L; C is of order U moves at constant velocity U. The flow rate through the pipe will be constant since L accommodates an integer number of spatial periods l of velocity perturbations (steps of the working cascade). This model encompasses the primary qualitative characteristics of the phenomenon of BPF oscillation generation at the blade machine outlet. In the case where U is significantly smaller than C, we observe quasi-stationary motion, with the velocity profile propagating along the pipe as desired. Conversely, if l is considerably larger than L, we witness oscillatory motion with a periodic change in the pipe's flow rate. In the considered case, in accordance with the postulate of cascading energy transfer from large vortices to more and more small ones, the initial vortex perturbations are rapidly damped by turbulent exchange, thereby generating acoustic perturbations that propagate further at the speed of sound. In the vicinity of the pipe entrance, a limited zone of vortex perturbations is formed, with a size that, according to the first law of turbulence, is of the order of l.

The representation of the oscillatory motion of a compressible fluid as a set of acoustic and vortex modes allows us to simplify the initial equations as if we were connecting two regions. In one of these regions, the motion is described by non-linear equations, while in the other, it is described by linear equations.

The convective transport of vortex perturbations can be considered the primary physical cause of the unsteady process of generating BPF pressure pulsations in the

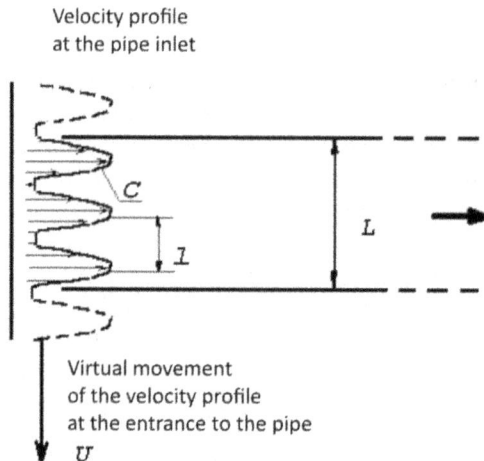

Velocity profile
at the pipe inlet

Virtual movement
of the velocity profile
at the entrance to the pipe

U

Figure 1.
Phenomenological model—semi-infinite pipe with vortex excitation of oscillations.

blade machine. These disturbances result from the movement of a periodically inho-mogeneous flow with circumferential velocity U of the impeller relative to the stator path.

In this context, the effects of viscosity and thermal phenomena are relatively insig-nificant, and thus, for the sake of simplicity, the corresponding terms are not included in the equations of motion. Furthermore, the flow is assumed to be isentropic.

4. Derivation of the acoustic-vortex equation

The Cauchy-Helmholtz theorem allows the introduction of the acoustic and vortex modes of motion of the working medium. The velocity of the compressible medium is represented as a vector sum of the main translational and rotational motions of the fluid as an incompressible medium (vortex mode) and small oscillation due to com-pressibility (acoustic mode).

The following assumptions are made:

The flow is subsonic, isentropic, and viscous dumping is not taken into consider-ation with regard to acoustic fluctuations. Acoustic oscillations (due to the compress-ibility of the medium) are much smaller than vortex oscillations (pseudo-sound). The basic equations of motion of a compressible medium are as follows:

$$\frac{\partial \mathbf{V}}{\partial t} + \nabla \frac{V^2}{2} - \mathbf{V} \times (\nabla \times \mathbf{V}) = -\frac{\nabla P}{\rho} + \nu \Delta \mathbf{V} \tag{1}$$

$$\frac{\partial \rho}{\partial t} + \nabla(\rho \mathbf{V}) = 0 \tag{2}$$

$$s = const \tag{3}$$

For isentropic flow, the enthalpy, pressure, and density increments are related by thermodynamic relations.

$$di = \frac{dP}{\rho}, dP = c_0^2 d\rho, \tag{4}$$

where c_0—is the speed of sound in the undisturbed working medium.

Taking into account relations (Eq. (4)), we rewrite (Eqs. (1) and (2)) in a form suitable for further transformations:

$$\frac{\partial \mathbf{V}}{\partial t} + \nabla \frac{V^2}{2} - \mathbf{V} \times (\nabla \times \mathbf{V}) = -\nabla i + \nu \Delta \mathbf{V} \tag{5}$$

$$\frac{1}{c_0^2} \left(\frac{\partial i}{\partial t} + (\nabla \mathbf{V})i \right) + \nabla \mathbf{V} = 0 \tag{6}$$

Based on the Cauchy-Helmholtz theorem, the unsteady fluid velocity can be defined as the sum of the vortex mode velocity $\mathbf{U}$ (translational and rotational motion of a completely incompressible medium) and the acoustic motion velocity $\mathbf{V}_a$.

Let us introduce a scalar function—the acoustic potential φ. Then, the acoustic velocity

$$\mathbf{V}_a = \nabla\varphi \qquad (7)$$

Thus, for the velocity of a compressible medium, we obtain the following expression:

$$\mathbf{V} = \mathbf{U} + \nabla\varphi = \mathbf{U} + V_a \qquad (8)$$

We will consider a subsonic flow $M = U/c_0 \ll 1$ with small acoustic oscillations ($V_a \ll c_0$). Let us also write down the following obvious relations:

$$\nabla\mathbf{U} = 0, \nabla \times \mathbf{V} = \nabla \times \mathbf{U} \qquad (9)$$

Since

$$\nabla \times \mathbf{V}_a \equiv \nabla \times \nabla\varphi \equiv 0 \qquad (10)$$

Thus, the vortex perturbations of the flow are determined by the velocity of the incompressible (divergence-free) flow. Let us now substitute relation (Eq. (8)) into (Eq. (5)). After simple transformations of (Eq. (5)) we obtain.

$$\frac{d\mathbf{U}}{dt} = -\nabla H + \nu\Delta\mathbf{U} + \nabla\varphi \times \nabla \times \mathbf{U}, \qquad (11)$$

where

$$H = i + \frac{d\varphi}{dt} + \frac{1}{2}(\nabla\varphi)^2 \qquad (12)$$

$$\frac{d}{dt} = \frac{\partial}{\partial t} + \mathbf{U}\nabla \qquad (13)$$

The term $\nabla\varphi \times \nabla \times \mathbf{U}$ in (Eq. (11)) reflects the interaction of acoustic and vortex modes. It is important to note that this term can be significant in the acoustic resonance zone and the formation of feedback between acoustic waves and the mechanism of instability waves in the mixing layer or in the formation of concentrated vortices, which in turn generate acoustic oscillations. In this case, this term can be neglected.

Concurrently, Eq. (11) incorporates the influence of viscous forces, which contribute to the uneven distribution of flow parameters along the rotor blade pitch and the generation of pressure oscillations at the blade frequencies.

Substituting formula Eq. (12) into Eq. (6) and dividing the velocity into acoustic and vortex modes simultaneously yields the following result:

Now, by substituting i from formula Eq. (12) into Eq. (6) and simultaneously splitting the velocity $\mathbf{V}$ into acoustic and vortex modes, we obtain

$$\frac{1}{c_0^2}\frac{d}{dt}\left(\frac{d\varphi}{dt} + \frac{(\nabla\varphi)^2}{2}\right) - \Delta\varphi + \frac{1}{c_0^2}\nabla\varphi\nabla\left(\frac{d\varphi}{dt} + \frac{(\nabla\varphi)^2}{2}\right) = \frac{1}{c_0^2}\left(\frac{dH}{dt} + \nabla\varphi\nabla H\right) \qquad (14)$$

Taking into account the aforementioned assumptions, Eq. (14) can be linearised with respect to φ, resulting in the following form:

$$\frac{1}{c_0^2}\frac{d^2\varphi}{dt^2} - \Delta\varphi = \frac{1}{c_0^2}\frac{dH}{dt} \tag{15}$$

From Eqs. (14) and (15), the unsteady vortex motion of the fluid generates acoustic oscillations. As a source of acoustic oscillations in the inhomogeneous wave-acoustic equation, the term in the right-hand part of the equation is $\frac{1}{c_0^2}\frac{dH}{dt}$.

Taking into account the last assumption, linearisation by φ, and neglecting the convection of the acoustic mode, relations (Eqs. (11) and (12)) are written in the following form:

$$\frac{d\mathbf{U}}{dt} = -\nabla H + \nu\Delta\mathbf{U} \tag{16}$$

$$-\frac{\partial\varphi}{\partial t} = i - H \tag{17}$$

Eqs. (16) and (17) provide a solution to the problem of splitting the fundamental equations of motion of a compressible medium into vortex and acoustic modes. Eq. (16) describes the vortex turbulent motion of an incompressible viscous fluid under the action of an unsteady pressure gradient $\nabla P_v = \rho_0\nabla H$.

We differentiate Eq. (15) by taking the partial derivative in time and replace it with the expression for $\frac{\partial\varphi}{\partial t}$ from Eq. (17). After simple transformations, we obtain the following:

$$\frac{1}{c_0^2}\frac{\partial^2 i}{\partial t^2} - \Delta i = -\Delta H \tag{18}$$

The source function in the right part of equation (Eq. (18)) is calculated from the vortex mode velocity field after solving equation (Eq. (16)):

$$-\Delta H = \nabla \cdot (\mathbf{U} \cdot \nabla\mathbf{U}) = \nabla \cdot \left(\nabla\left(\frac{1}{2}U^2\right) - \mathbf{U} \times \nabla \times \mathbf{U}\right) \tag{19}$$

By neglecting the convective terms in the time derivative of equation (Eq. (18)), we obtain:

$$\frac{1}{c_0^2}\frac{\partial^2 i}{\partial t^2} - \Delta i = s \tag{20}$$

The ripple part of the function S delivered from the velocity field of the vortex mode by relation (Eq. (21)) is denoted by accounting (Eq. (22)). Thus, $s = S - S_0$ is the source function of the wave equation (Eq. (2)).

$$S = \nabla\left(\nabla\left(\frac{1}{2}U^2\right) - \mathbf{U} \times \nabla \times \mathbf{U}\right) \tag{21}$$

For an undisturbed medium, the acoustic potential $\varphi = 0$, thus

$$i_0 = H_0 \ ; \ S_0 = -\Delta H_0 \tag{22}$$

The functions H, s can be expressed through mean values and ripple components as follows

$$h = i - i_0; \quad g = H - H_0 \tag{23}$$

The amplitude of pressure pulsations is two or three orders of magnitude lower than the mean pressure. Therefore, for enthalpy fluctuations, we can approximate the following

$$h \approx \frac{(P - P_0)}{\rho_0} = \frac{P'}{\rho_0} \tag{24}$$

Similarly, for vortex mode oscillations (pseudo-sound), we have

$$g \approx \frac{H - H_0}{\rho_0} = \frac{P'_v}{\rho_0}, \tag{25}$$

where P'—sound pressure, P_v'—pseudo-sound oscillation.
It is necessary to consider Eqs. (17) and (26).

$$P' = P'_v - \rho_0 \frac{\partial \varphi}{\partial t} \tag{26}$$

It is necessary to solve the obtained Eqs. (16) and (20) independently, taking into account the corresponding boundary conditions:

$$\left. \frac{\partial h}{\partial n} \right|^m = -\frac{1}{c_0} \frac{1}{Z} \left(\frac{h^{m+1} - h^{m-1}}{2\Delta t} \right) + \frac{\partial g}{\partial n} + \frac{1}{c_0} \frac{1}{Z} \frac{\partial g}{\partial t} \tag{27}$$

By solving the wave equation by explicit method (m-is the time level index) at the impedance boundary with normal incidence n and taking into account the pseudo-sound oscillations (if necessary), we can determine the derivative along the normal to the boundary surface through the time derivative and the specific complex acoustic impedance Z for the BPF harmonic. Pseudo-sound oscillations are known from the solution for the vortex mode.

5. Some results about pseudo-sound and acoustic oscillations

An important advantage of the acoustic-vortex method is that it provides simultaneous results about vortex mode pressure fluctuations (pseudo-sound) and acoustic pulsations, which follow from Eq. (26).

Such information makes it possible to optimise the acoustic far-field and vibration loads of the structure based on a single solution.

Figure 2 presents a comparison of pressure pulsation measurements on a two-dimensional model of an air centrifugal pump [66] with a calculation by different methods.

It is evident that in a narrow zone near the outlet of the centrifugal impeller, a high amplitude of pressure pulsations is recorded due to vortex oscillations, which decay rapidly, leaving acoustic pulsations propagating towards the open outlet of the model.

Another example demonstrates a comparison of the calculated and measured amplitude of the first BPF harmonic in the inlet duct for an axial fan with a blade radius of R = 60 mm.

Total amplitude of pressure pulsation

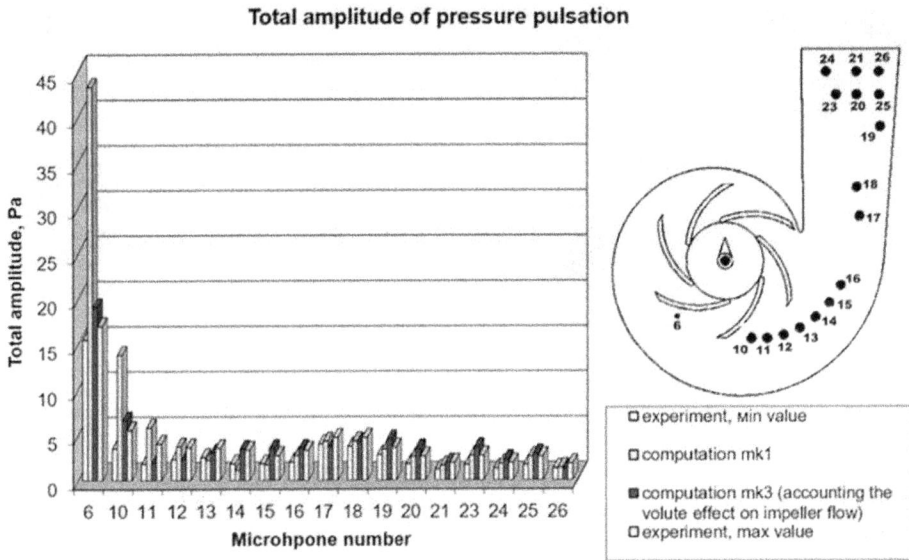

Figure 2.
Comparison of pressure pulsations in the volute of model pump.

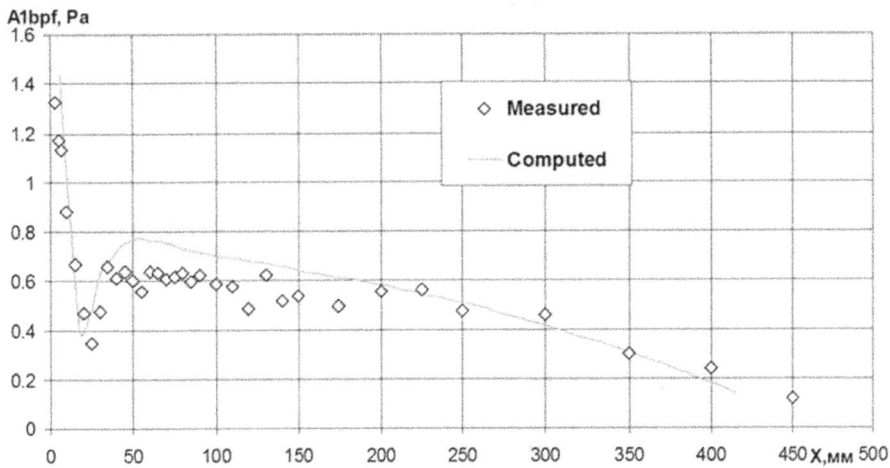

Figure 3.
Amplitude of the first BPF harmonic on a radius of 1.22R upstream from rotor.

Good agreement is obtained at a radius of 1.22R (**Figure 3**), where the BPF amplitude near the rotor is at a relatively low level. The curve exhibits a non-monotonic character due to the interaction of pseudo-sound oscillations and acoustic waves reflected from the open end of the duct [67].

Figure 4 provides an additional example of the instantaneous spatial configuration of the source function for the first BPF harmonic, calculated from the vortex mode velocity field for a 70-mm-diameter quadcopter propeller. This example offers further

Figure 4.
Instantaneous spatial configuration of the source function S for the first BPF harmonic of the quadcopter two-blades propeller.

insight into the process of pressure pulsations generation and possible ways of optimisation.

6. Conclusion

Application of the acoustic-vortex decomposition method represents a significant advancement in the field of blade machine acoustics. By decomposing the velocity field of the compressible medium into distinct vortex and acoustic modes, the method offers a robust framework for understanding and quantifying pressure pulsations and noise emission.

In the acoustic-vortex method, the noise source is represented as a local derivative of the vortex mode pressure. This defines the source of pressure pulsations and near-field noise as a direct result of numerical modelling.

Mathematically rigorous derivation of an inhomogeneous wave equation governing enthalpy pulsations, which integrates pseudo-sound oscillations within the source vortex region and acoustic oscillations in the near-field and far-field domains, ensures a comprehensive treatment of noise generation mechanisms but also addresses complexities such as flow inhomogeneity, turbulence effects, rotor interference, and surface impedance characteristics. One of the method's strengths is its formulation of boundary conditions using complex specific acoustic impedance and pseudo-sound oscillations, which enhances accuracy in predicting noise propagation dynamics.

This method is widely used to analyse the tonal noise of centrifugal fans and pumps in the two-dimensional approximation and has been experimentally verified in detail on a centrifugal pump and axial ventilator model. A three-dimensional version of the acoustic-vortex method has been developed for axial centrifugal pumps, axial fans, and multicopter propellers.

The principal advantage of this method is that it is not constrained by the conventions of aeroacoustic analogy, thereby enabling greater accuracy in optimisation work. By moving away from conventional aeroacoustic analogies, it establishes a direct link between vortex mode velocity fields and noise sources, thereby reducing arbitrariness in modelling outcomes. This methodological shift not only improves computational efficiency but also enhances the reliability of numerical simulations in predicting noise levels in blade machines.

Nomenclature

$\mathbf{V}$	velocity of compressible medium
P	pressure
ρ	density
ν	kinematic viscosity
s	entropy
i	enthalpy
c_0	speed of sound
$\mathbf{U}$	velocity of the vortex mode
$\mathbf{V}_a$	acoustic velocity
φ	acoustic potential
H	enthalpy in the vortex mode flow
P_v	pressure in the vortex mode flow
f	ripple source function
F	source function
h	pulsations of enthalpy
g	pseudo-sound pulsations
Z	specific complex acoustic impedance

Indexes

0	undisturbed medium
$'$	pulsatory part

Author details

Sergey Timushev[1]*, Andrey Aksenov[2] and Jiawen Li[3]

1 Moscow Aviation Institute, Moscow, Russia

2 Tesis Ltd., Moscow, Russia

3 School of Astronautics Beihang University (BUAA), Beijing, China

*Address all correspondence to: irico.harmony@gmail.com

IntechOpen

References

[1] ISO 3740:2019(en) Acoustics — Determination of sound power levels of noise sources — Guidelines for the use of basic standards

[2] ISO 11690-1:2020(en) Acoustics — Recommended practice for the design of low-noise workplaces containing machinery — Part 1: Noise control strategies

[3] Aksenov AA, Gavrilyuk VN, Timushev SF. Numerical simulation of tonal fan noise of computers and air conditioning systems. Acoustical Physics. 2016;**62**(4):447-455

[4] Jiang Y, Åbom M, Feng L, Timouchev S, Maliczak C. Investigation of sound generation from an axial fan for engine cooling. In: Proceedings of 3rd International Symposium on fan Noise 2007, 19–21 September, 2007, Lyon, France

[5] Chen YN. Water-pressure oscillations in the volute casings of storage pumps. In: Sulzer Technical Review Research Issue (Turbo-Machinery). 1961. pp. 21-34

[6] Shtrub RA. Pressure fluctuations and fatigue stresses in storage pumps and pump turbines. Journal of Engineering for Power. 1964;**86**(2):191-194. DOI: 10.1115/1.3677575

[7] Burnside N, Roy T. Whining Drones Bringing Burritos and Coffee are Bitterly Dividing Canberra Residents. ABCNews; 2018. Available from: https://www.abc.net.au/news/2018-11-09/noise-from-dronedelivery-service-divides-canberra-residents/10484044 [Accessed: August 14, 2024]

[8] BBC News: Why your Pizza May Never be Delivered by Drone. Available from: https://www.bbc.com/news/business-46483178 [Accessed: March 11, 2019]

[9] Moshkov PA, Samokhin VF, Yakovlev AA. Selection of an audibility criterion for propeller driven unmanned aerial vehicle. Russian Aeronautics. 2018;**61**(2):149-155. DOI: 10.3103/S1068799818020010

[10] Moshkov P, Ostrikov N, Samokhin V, Valiev A. Study of Ptero-G0 UAV noise with level flight conditions. In: Proceedings of the 25th AIAA/CEAS Aeroacoustics Conference. 2019. AIAA Paper No. 2019-2514. DOI: 10.2514/6.2019-2514

[11] Franssen E, Wichen VC, Nagelkerke N, Lebret E. Aircraft noise around a large international airport and its impact on general health and medication use. Occupational and Environmental Medicine. 2004;**61**(5):405-413

[12] Swift H. A review of the literature related to potential health effects of aircraft noise-PARTNERCOE-2010-003. In: Partnership for Air Transportation Noise and Emissions Reduction. 2010

[13] Sinibaldi G, Marino L. Experimental analysis on the noise of propellers for small UAV. Applied Acoustics. 2013;**74**(1):79-88

[14] Moshkov P, Samokhin V, Yakovlev A, Bolun C. The problems of selecting the power plant for light propeller-driven aircraft and unmanned aerial vehicle taking into account the requirements for community noise. Akustika. 2021;**39**:164-169

[15] Guelich JF, Bolleter U. Pressure pulsations in centrifugal pumps. Transactions of the ASME. Journal of Vibration and Acoustics. 1992;**114**:272-279

[16] Petit O, Nilsson H. Numerical investigations of unsteady flow in a centrifugal pump with a vaned diffuser. International Journal of Rotating Machinery. 2013;**2013**(961580):1-14. DOI: 10.1155/2013/961580

[17] Zogg UB. Generation and propagation of hydraulic noise in centrifugal pumps. Pump noise and vibration. In: Proc. 1st Int. Symp. Pump Noise and Vibration (Clamart, France 7–9 July 1993). Senlis (France): CETIM; 1993. pp. 263-270

[18] Solomakhova TS. Radial Fans. Moscow: Nauka; 2015. 460p [in Russian]

[19] Zotov BN. Investigation of hydrodynamic vibration sources of a centrifugal pump. Power Engineering. 1974;(2):26-32. [in Russian]

[20] Offenhäuser H. Druckschwankungsmessungen an Kreiselpumpen mit Leitrad. VDI – Berichte Nr.193; 1973. p. 211 [in German]

[21] Jiang YY, Yoshimura S, Imai R, Katsura H, Yoshida T, Kato C. Quantitative evaluation of flow-induced structural vibration and noise in turbomachinery by full-scale weakly coupled simulation. Journal of Fluids and Structures. 2007;**23**:531-544

[22] Tinney CE, Sirohi J. Multirotor drone noise at static thrust. AIAA Journal. 2018;**56**:2816-2826

[23] Intaratep N, Alexander WN, Devenport WJ, Grace SM, Dropkin A. Experimental study of quadcopter acoustics and performance at static thrust conditions. In: 22nd AIAA/CEAS Proceedings of Aeroacoustics Conference. June 2016. DOI: 10.2514/6.2016-2873

[24] Pokrovsky BV, Yudin EY. The main features of noise and vibration of centrifugal pumps. Acoustic Journal. 1966;**XII**(3):355-364. [in Russian]

[25] Borovsky BI, Ershov NS, Ovsyannikov BV, Petrov VI, Chebaevsky VF, Shapiro AS. High-speed blade pumps. In: Ovsyannikova BV, Chebaevsky VF, editors. Moscow: Mashinostroenie; 1975. [in Russian]

[26] Selifonov VS, Ovsyannikov BV. Cavitation self-oscillations in pumps. Bulletin of MAI. 1995;(1). [in Russian]

[27] Pokrovsky BV, Ya RV. Cavitation in the outlet and its effect on the vibration of a centrifugal pump. In: Proceedings of "VNIIgidromash", Moscow. No. 45. 1974. [in Russian]

[28] Pokrovsky BV. Cavitation noise and vibrations of centrifugal pumps. In: Proceedings of "VNIIgidromash", Moscow. No. 39. 1969. [in Russian]

[29] Khoroshev GA. Pump vibrations caused by cavitation. Energo-mashinostroenie. 1960;(4):1. [in Russian]

[30] Kozelkov VP, Ya ZV, Efimochkin AF. Visual investigation of a cavitating centrifugal pump. In: Hydrodynamics of Blade Machines and General Mechanics. Voronezh: Voronezh Polytechnic Institute; 1977. [in Russian]

[31] Pokrovsky BV, Ya RV. On the reduction of discrete components from the inhomogeneity of the flow in the vibration spectrum of a centrifugal pump. In: Proceedings of the Seminar-Meeting "Issues of Vibration and Noise Damping in Structures and Machines", Kirov. 1970. [in Russian]

[32] Zotov BN. Vibration at blade frequencies in centrifugal pumps with the same number of wheel blades and

guied vanes. In: Gryanko LP, Papir AN, editors. Vane Pumps. Leningrad: Mashinostroenie; 1975. [in Russian]

[33] Croba D, Kueny JL. Unsteady flow computation in a centrifugal pump coupling of the impeller and the volute. In: Noise F, editor. An International INCE Symposium, Senlis (France). Proceedings. CETIM; 1992

[34] Croba D, Kueny JL, Hureau F, Kermarec J. Numerical and experimental unsteady flow analysis in centrifugal pumps. Impeller and volute interaction. Pump noise and vibrations. In: Proceedings 1st International Symposium. Senlis (France): CETIM; 1993

[35] Chu S, Dong R, Katz J. The effect of blade-tongue interactions on the flow structure, pressure fluctuations and noise within a centrifugal pump. Pump noise and vibrations. In: Proceedings 1st International Symposium. Senlis (France): CETIM; 1993

[36] Thompson MC, Hourigan K, Stokes AN. Prediction of the noise generation in a centrifugal fan by solution of the acoustic wave equation. In: Noise F, editor. An International INCE Symposium, Senlis (France). Proceedings. CETIM; 1992

[37] Howe MS. Contribution to the theory of aerodynamic sound, with application to excess jet noise and the theory of the flute. Journal of Fluid Mechanics. 1975;71(4):625-673

[38] Lighthill MJ. On sound generated aerodynamically. Part I. General theory. Proceedings of the Royal Society, London A. 1952;211:564-587

[39] Curle N. The influence of solid boundaries upon aerodynamic sound. Proceedings of the Royal Society, A. 1955;231:505-514

[40] Ffowcs-Williams JE, Hawkings DL. Sound generation by turbulence and surfaces in arbitrary motion. Philosophical Transactions of the Royal Society A. 1969;264:321-342

[41] Farassat F, Myers MK. Extension of Kirchhhoff's formula to radiation from moving surfaces. Journal of Sound and Vibration. 1988;123:451-461

[42] Liu H-l, Dai H-w, Ding J, Tan M-g, Wang Y, Huang H-q. Numerical and experimental studies of hydraulic noise induced by surface dipole sources in a centrifugal pump. Journal of Hydrodynamics. 2016;28(1):43-51. DOI: 10.1016/S1001-6058 (16)60606-6

[43] Sears WR. Some aspects of non-stationary airfoil theory and its practical application. Journal of the Aeronautical Sciences. 1941;8(3):104-108

[44] Atassi H, Hamad G. Sound generated in a cascade by three-dimensional disturbances convected in a subsonic flow/AIAA-81-2046. In: 7th AIAA Aero-Acoustics Conference. Palo Alto, California. 5-7 October; 1981

[45] Tyler JM, Sofrin TG. Axial flow compressor noise studies. SAE Transactions. 1962;70:309-332

[46] Caro S, Moreau S. Comparaison d'une technique 2D de type Sears avec un calcul instationnaire direct pour le calcul du bruit de raies d'un ventilateur. Bruit des ventilateurs à basse vitesse. Actes du colloque tenu à l'Ecole Centrale de Lyon les 8 et 9 novembre 2001. [in French]

[47] Caro S, Sandboge R, Iyer J, Nishio Y. Presentation of a CAA formulation based on Lighthill's analogy for fan noise. In: Proceedings of 3rd International Symposium on fan Noise 2007, Lyon, France, 19–21 September, 2007

[48] Sandboge R, Washburn K, Peak C. Validation of a CAA formulation based on Lighthill's analogy for a cooling fan and Mower blade noise. In: Proceedings of 3rd International Symposium on fan Noise 2007, Lyon, France, 19–21 September, 2007

[49] Zhu YJ, Ou YH, Tian J. Experimental and numerical investigation on noise of rotor blade passing outlet grille. NCEJ. 2008

[50] Silouane D, R, Nicolas Z, Andrew H. Use OpenFOAM coupled with finite and boundary element formulations for computational aero-acoustics for ducted obstacles. Inter Noise. 2019

[51] Ribner HS. The generation of sound by turbulent jets. Advances in Applied Mechanics. 1964;8:103-182

[52] Guédel A. Acoustique des ventilateurs. CETIAT. France: PYC LIVRES; 1999. [in French]

[53] Frost W, Moulden TH, editors. Handbook of Turbulence. Fundamentals and Applications. New York and London: Plenum Press; 1977

[54] Fraser WN, Kasassik IJ, Bush AR. Study of pump pulsation, surge and vibration throws light on reliability vs efficiency. Power. 1977

[55] Fraser WN. Recirculation in Centrifugal Pumps. World Pumps; 1982-188. pp. 227-235

[56] Sazonov AA. Investigation of some non-stationary phenomena in centrifugal pumps. In: Blade Machines and Jet Apparatuses. No. 6. Moscow: Mashinostroenie; 1972. [in Russian]

[57] Lokshin IL. Investigation of the flow behind the wheels of centrifugal fans in relative motion. In: Industrial

Aerodynamics/TsAGI. No. 12. Moscow: Oborongiz; 1959. [in Russian]

[58] Raj D, Swim WB. Measurements of the mean flow velocity and velocity fluctuations at the exit of an FC centrifugal fan rotor. ASME. Journal of Engineering and Power. 1981;103(2): 393-399. DOI: 10.1115/1.3230733

[59] Seleznev KP, Galerkin YB. Centrifugal Compressors. Leningrad: Mashinostroenie; 1982. [in Russian]

[60] Eckardt D. Instantaneous measurements in the jet-wake discharge flow of a centrifugal compressor impeller. Journal of Engineering and Power. 1975;97(3):337-345. DOI: 10.1115/1.3445999

[61] Johnson MW, Moore J. The development of wake flow in a centrifugal impeller. Journal of Engineering and Power. 1980;102(2): 382-389. DOI: 10.1115/1.3230265

[62] Timshin AI. Experimental study of the flow structure at the outlet of the centrifugal pump wheel. In: Hydraulic Machines. Kharkiv: Publishing House of KHSU; 1971. [in Russian]

[63] Erdreich VS, Zhukovsky RN. Investigation of flow unsteadiness in a model radial-axial pump turbine in pumping mode. In: Proceedings of "VNIIgidromash", Moscow. No. 44. 1972. [in Russian]

[64] Johnston JP, Halleen RM, Lezius DK. Effects of spanwise rotation on the structure of two-dimensional fully developed turbulent channel flow. Journal of Fluid Mechanics. 1972;56(Part 3):533

[65] Wagner RE, Velkoff HR. Measurement of secondary flows in a

rotating duct. Journal of Engineering for
Power. 1972;**94**(4):261-270.
DOI: 10.1115/1.3445681

[66] Timouchev S, Tourret J, Pavic G,
Aksenov A. Numerical 2-D and 3-D
methods for computation of internal
unsteady pressure field and sound near
field of fans. In: 2nd International
Symposium fan Noise 2003, Senlis
(France), 23–25 September 2003 /
Proceedings

[67] Timouchev S, Nedashkovsky A,
Pavic G. Experimental validation of axial
fan 3D acoustic-vortex method CFD-
CAA study. In: Proceedings of 3rd
International Symposium on fan Noise
2007; 19–21 September, 2007; Lyon,
France

Chapter 5

Thermal Effect on Taylor-Couette Flow Dynamics

Hayato Masuda and Naoto Ohmura

Abstract

In this chapter, we experimentally and numerically reviewed Taylor-Couette flow dynamics with an axial temperature distribution. In experiments, the glycerol aqueous solution with various concentrations was used. Flow pattern observation and temperature measurement were performed. Based on the results, we classified the flow pattern into six cases as a function of Reynolds and Grashof numbers. In the specified case (denoted as Case II in this chapter), the Taylor vortex flow and heat convection alternatively appeared. In this condition, when switching from the Taylor vortex flow to heat convection, heat/mass was rapidly transferred. Using numerical simulation, the fluid flow and heat transfer in Case II were investigated. As a result, an average Nusselt number in Case II was quite large compared with that in the stable Taylor vortex flow regime (higher Re case). Therefore, heat transfer augmentation with lower power input is expected if the interaction between the Taylor vortex flow and heat transfer is applied.

Keywords: Taylor-Couette flow, thermo-fluid dynamics, heat convection, heat transfer, flow pattern, numerical simulation

1. Introduction

In the flow between coaxial cylinders with the inner one rotating, a Taylor vortex flow appears above a critical Reynolds number, Re_{cr}, as shown in **Figure 1**, [1]. This flow is generally called Taylor-Couette flow. The Taylor-Couette flow exhibits a cascade type of flow transition from laminar to fully-developed turbulent flows, experiencing several stable states: (i) laminar Taylor vortex flow, (ii) singly periodic wavy vortex flow, (iii) quasi-periodic wavy vortex flow, and (iv) weakly turbulent wavy vortex flow [2]. The Taylor vortex flow has been attracting attention from a various perspective, e.g., mathematics [3], fluid mechanics [4], and process engineering [5]. In particular, the Taylor vortex flow has several advantages for process engineers. The mixing within each cell, called Taylor cells, is promoted due to the toroidal motion [6]. At the stagnation point on the outer cylinder surface, heat and mass transfer are enhanced due to the outward jet [7]. In addition, Taylor cells through in a single file without breaking when the relatively small axial flow is imposed [5]. Thus, continuous processing is possible

(a)   (b)

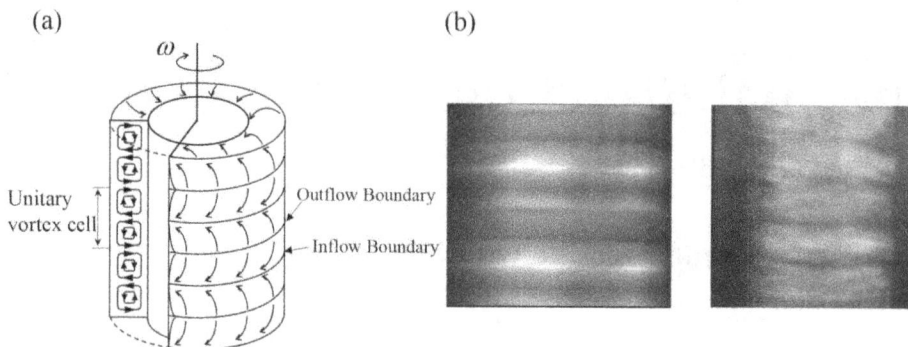

Figure 1.
Taylor-Couette flow: (a) Schematic picture and (b) flow visualization [1]. The left and right figures show the laminar Taylor vortex flow and wavy vortex flow, respectively. The black band corresponds to inflow boundaries.

while mixing and heat/mass transfer are enhanced. So far, the Taylor-Couette flow has been applied to various processes, e.g., polymerization reaction [8], particle synthesis [9], enzymatic reaction [10, 11], and particle separation [12].

When applying Taylor-Couette flow to processes, thermal operation is often necessary to promote heating/cooling or chemical reactions. As a result, the temperature distribution is created. If the viscous and inertial forces are not relatively strong compared with the buoyant forces, heat convection occurs, driven by buoyancy. Heat convection disturbs the stability of Taylor-Couette flow; consequently, new types of flow patterns were observed depending on Re and Grashof number, Gr. Ball et al. [13] reported that a spiral vortex structure is formed in a vertical Taylor-Couette flow with a radial temperature difference. Kuo and Ball [14] numerically showed the mechanism of the spiral motion and the deformation of Taylor cells in vertical Taylor-Couette flow having a radial temperature difference. As another example, in highly viscous fluid systems, the temperature distribution is generated by viscous heating, leading to complicated flow patterns [15–17].

From the perspective of process engineering, not only radial but also axial temperature distributions are an important topic. For example, exothermic reaction at the initial stage of reaction occurs in the bulk polymerization reaction of styrene [18]. As with other examples, in starch hydrolysis consisting of two steps (gelatinization and saccharification), the temperature should be step-wisely changed because the suitable temperature for each step significantly differed [10, 11]. If the continuous Taylor-Couette flow with axial flow is applied to such processes, the axial temperature distribution would be neglected. Recently, Leng et al. [19] numerically investigated heat transfer and fluid flow with axial temperature gradient. Although their study employed a short column (i.e., low aspect ratio), the longer column should be utilized for practical investigation.

We have experimentally and numerically investigated the thermo-fluid dynamics of Taylor-Couette flow with an axial temperature distribution for a moderate aspect ratio [20, 21]. The target of our study was within the moderate Re regime because of the assumption for utilization in chemical processes with highly viscous fluids. In the Re regime, unique flow patterns were observed under specified conditions, leading to the enhancement of heat/mass transfer. In this chapter, we review the fluid flow and heat/mass transfer characteristics varying parameters (physical properties of the fluid, Re, Gr, and degree of axial temperature distribution).

2. Experiment

The apparatus used in our study is depicted in **Figure 2**. The Taylor-Couette flow apparatus comprised two concentric cylinders (the length L: 195 mm) and water jackets. The radii of the inner and outer cylinders are R_i = 25 mm and R_o = 38 mm. To create the axial temperature distribution, the water jacket was divided into two parts. The temperatures of jackets for lower (T_L) and upper (T_U) were varied with limitation $T_L > T_U$. Due to this condition, buoyancy is generated, consequently, the interaction between Taylor-Couette flow and heat convection can be investigated. The temperature at the boundary of the upper and lower jackets was measured. This temperature was denoted as T_b [21] or T_m [20]. In addition, the temperatures at 40 mm from above and below the measuring point for T_b (or T_m) were also measured. Each temperature was referred to as T_{bu} and T_{bl} [21] or T_u and T_l [20]. The glycerol aqueous solution having various concentrations (C = 40–90 wt%) was used for experiments. With respect to the physical properties of the fluid, the density and viscosity were measured by a density meter (DA-640, Kyoto Electronics Manufacturing Co., Ltd.) and a rheometer (MCR102, Anton Paar GmbH).

The thermo-fluid dynamics is characterized by *Re* and *Gr*, which are defined as:

$$\text{Re} = \frac{\rho \omega R_i d}{\eta} \tag{1}$$

$$\text{Gr} = \frac{g \rho^2 \beta (T_L - T_U) d^3}{\eta^2} \tag{2}$$

where ρ is the density, ω is the rotational velocity of the inner cylinder, d is the gap width, η is the viscosity, g is the gravitational acceleration, β is the coefficient of volume expansion. The density and viscosity at $(T_L + T_U)/2$ were used as the representative values. The coefficient of volume expansion was estimated from the published database [22].

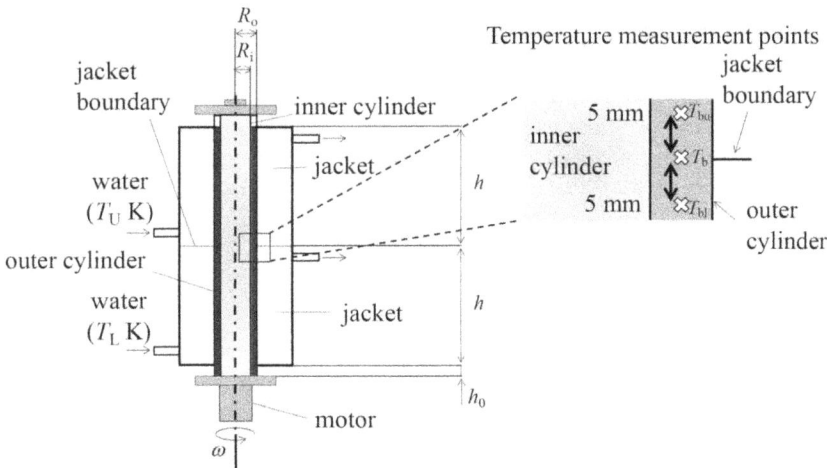

Figure 2.
Experimental apparatus [21].

3. Numerical simulation

To investigate the detailed characteristics of fluid flow and heat transfer, we employed numerical simulation. The governing equations are (3) continuity, (4) Navier–Stokes, and (5) energy conservation equations, as follows:

$$\nabla \cdot \mathbf{u} = 0 \tag{3}$$

$$\frac{\partial \mathbf{u}}{\partial t} + (\mathbf{u} \cdot \nabla)\mathbf{u} = -\frac{\nabla p}{\rho} + \frac{1}{\rho}\nabla \cdot (2\eta \mathbf{D}) - \mathbf{g}\beta(T - T_{ref}) \tag{4}$$

$$\frac{\partial T}{\partial t} + \nabla \cdot (T\mathbf{u}) = \alpha \nabla^2 T \tag{5}$$

where $\mathbf{u}$ is the velocity, t is the time, p is the pressure, $\mathbf{D}$ is the deformation rate tensor, T is the temperature, T_{ref} is the reference temperature, and α is the thermal diffusivity. The Boussinesq approximation term was added to the Navier–Stokes equation. In the simulation, the fluid was assumed to be a 40 wt% glycerol aqueous solution. To take the effect of temperature on the viscosity into consideration, an Arrhenius-type equation was used, as follows:

$$\eta = \eta_{ref}\left\{\frac{E}{R}\left(\frac{1}{T} - \frac{1}{T_{ref}}\right)\right\} \tag{6}$$

η_{ref} is the viscosity at the reference temperature (T_{ref}), E is the activation energy and R is the gas constant. The viscosities of 40 wt% glycerol aqueous solution at various T were measured by the rheometer. By fitting the measured data to Eq. (6), values of each parameter were obtained. For other thermophysical properties (α and β), the reference data were used [22]. These values used in this simulation are summarized in **Table 1**.

The computational domain is shown in **Figure 3**. The geometrical configuration was the same as that of the experimental apparatus. The inner cylinder surface was assumed the rotating wall having the circumferential velocity of $R_i{\cdot}\omega$. The outer cylinder surface and the bottom surface were regarded as stationary wall, while the slip condition was imposed on the top plate. For pressure conditions, no gradient condition was imposed on any surface. With respect to thermal conditions at the surface of the outer cylinder, the continuous distribution of axial temperature was imposed because the temperature varied not step-wise but continuously. The temperature was expressed as a function of axial position (z), as follows:

$$T(z) = \frac{T_B - T_A}{1 + \exp\{a(z - 2h)\}} + T_A \tag{7}$$

η_{ref} [Pa·s]	0.0037
E [kJ/mol]	2.177
α [m²/s]	1.28×10^{-7}
β [1/K]	4.30×10^{-4}

Table 1.
Thermos and physical properties of fluid used in simulation.

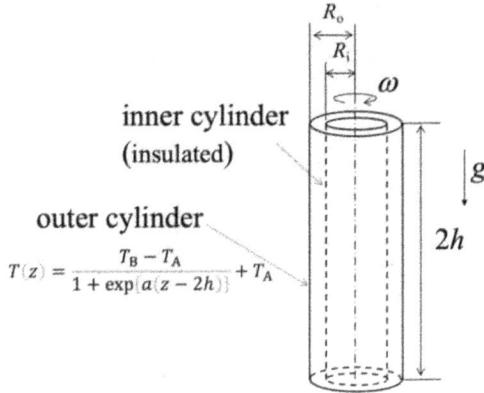

Figure 3.
Computational domain [21].

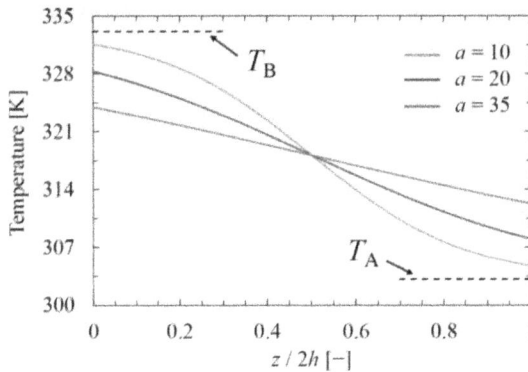

Figure 4.
Temperature profiles at the outer cylinder surface [21].

where a is an arbitrary parameter to control the gradient of temperature distribution in the axial direction. T_A and T_B were fixed at 303.15 K and 333.15 K in this simulation. For the value of a, a = 10, 20, and 35 were selected. The temperature profile at each a is shown in **Figure 4**. In this chapter, the results for a = 20 are presented.

In the simulation, the governing equations were discretized based on the finite volume method using OpenFOAM 4.1® code. The detailed procedure including the mesh resolution test was written in our previous paper [21].

4. Thermo-fluid dynamics in non-isothermal system

In the case of C = 40 wt%, the flow pattern was classified into three cases. When buoyancy is dominant, heat convection mainly appears without Taylor vortices. This case is denoted as Case I. As the centrifugal force increases, the Taylor vortex flow becomes dominant. When buoyancy is comparable to the centrifugal force, the Taylor vortex flow and heat convection alternately appear (Case II). When the centrifugal force is further increased, the stable Taylor vortex flow is generated (Case III). Typical examples of the flow pattern for each Case are shown in **Figure 5**. As shown in

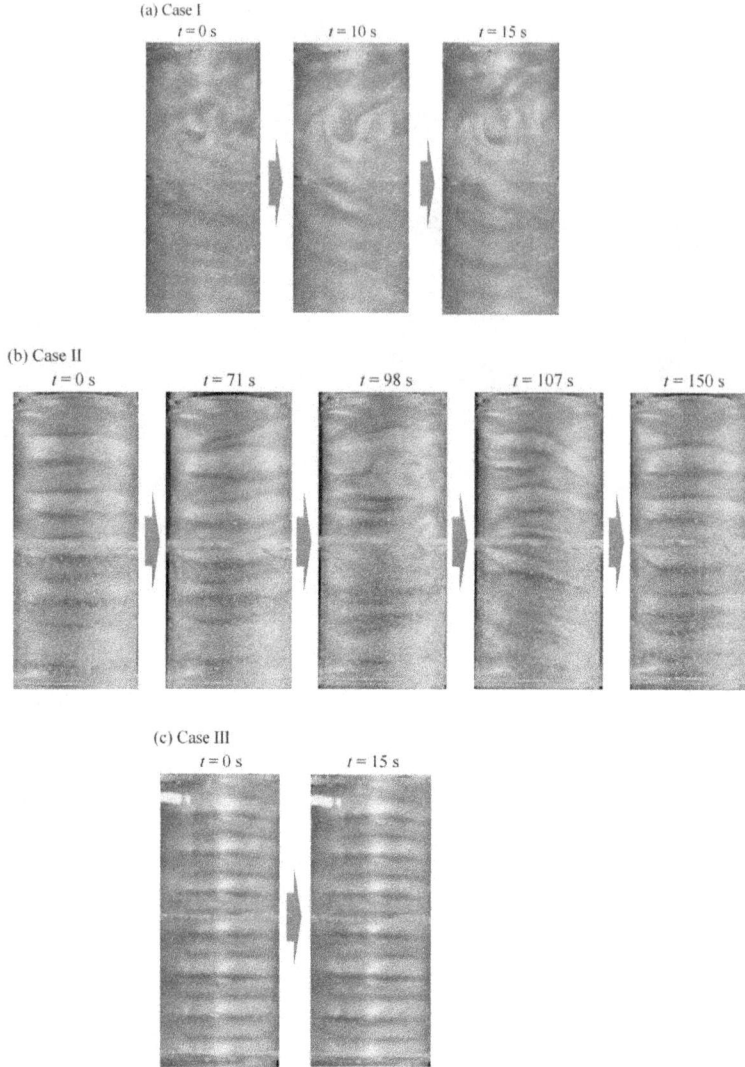

Figure 5.
Flow patterns at Gr = 8.8 × 10⁷: (a) Re = 124 (Case I), (b) Re = 496 (Case II), (c) Re = 620 (Case III) [20].

Figure 5a and **b**, it is confirmed that heat convection or Taylor vortex flow is dominant in Cases I and III. On the other hand, in Case II (**Figure 5b**), two types of flows dominate alternately. At t = 0 s, the stable Taylor vortex flow was observed. In this term, the axial temperature distribution in the field would not be crucial. The axial temperature distribution gradually develops because Taylor cells do not transport heat in the axial direction. As a result, the Taylor vortex flow began to fluctuate at t = 71 s due to the increased effect of buoyancy. After that, Taylor vortex flow collapsed (e.g., t = 98 and 107 s) and heat convection became dominant. When heat convection is dominant, the temperature distribution is uniformized due to the globally large loop. After the temperature distribution becomes small to suppress buoyancy, the Taylor vortex flow can be generated again. Thereafter, a series of fluctuations, collapse, and formation continued semi-permanently. The cross-sectional views for stable, deformed, and collapsed Taylor cells in Case II are shown in **Figure 6**. How to visualize using a laser sheet is written in our previous paper [20]. As shown in **Figure 6**, when the Taylor cells collapsed, many vortices that were generated by the

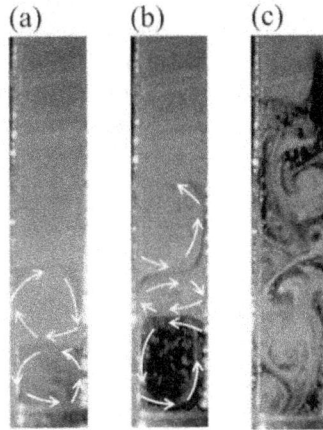

Figure 6.
Observed vortices at each stage in Case II (Re = 496, Gr = 8.8 × 10⁷): (a) stable Taylor vortices, (b) deformed Taylor vortices, and (c) collapsed Taylor vortices [20]. Arrows in figures denote the vortex motion.

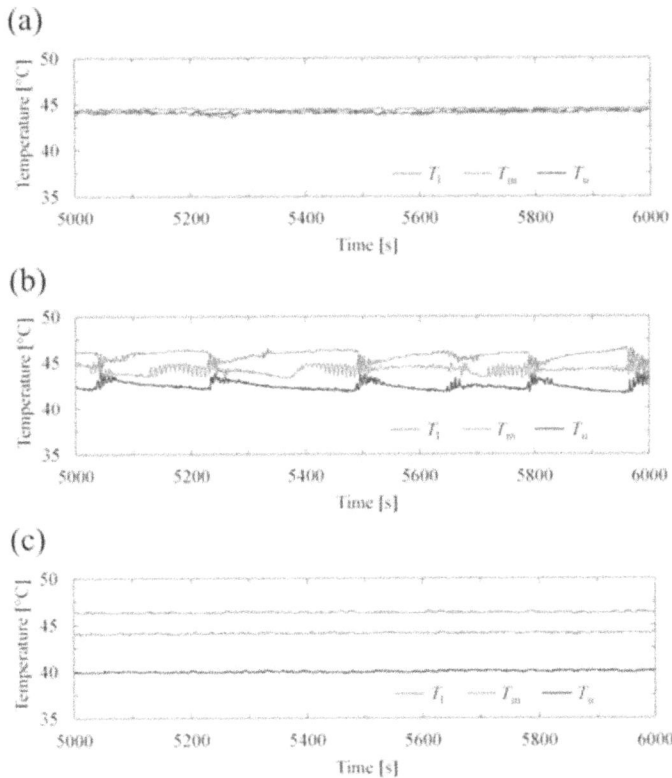

Figure 7.
Temperature (T_l, T_m, and T_u) change with respect to time after elapse of 5000 s from the start of the operation in each case: (a) Case I (Re = 124, Gr = 8.8 × 10⁷), (b) Case II (Re = 496, Gr = 8.8 × 10⁷), and (c) Case III (Re = 620, Gr = 8.8 × 10⁷) [20].

interaction between buoyancy and centrifugal forces significantly promoted global mixing, although the flow condition was laminar.

The temperature change with time in each case is shown in **Figure 7**. As shown in **Figure 7a**, there was no clear difference in axial temperature in Case I because of the large loop of heat convection. On the other hand, in Case III, the distinct distribution of axial temperature was found due to the stable Taylor vortex flow (**Figure 7c**). The unique temperature fluctuation was observed in Case II, as shown in **Figure 7b**. The temperature difference between T_l and T_u gradually increased although the Taylor vortices were formed. With the fluctuation of the Taylor vortices, the variation in Tm increased. When Taylor vortices were collapsed, the temperatures at each measurement point were closer to the same value. Therefore, the interaction between Taylor vortex flow and heat convection has a potential for heat transfer enhancement.

From a practical viewpoint, the thermo-fluid dynamics in other concentration systems remain unclear. For example, **Figure 8** shows the flow pattern and temperature change with time at C = 70 wt% (Re = 120.9 and Gr = 4591.2). In this condition, a new type of flow pattern, in which the Taylor cells fluctuate while maintaining the cellular structure was observed. Therefore, the temperature fluctuation with a distinct distribution in the axial direction (T_{bl}, T_b, and T_{bu}) was confirmed in **Figure 8b**. This flow mode is denoted as Case VI. At higher C, the other flow pattern was observed. **Figure 9** shows the flow pattern and temperature change with time at C = 80 wt% (Re = 48.6 and Gr = 516.3). In this condition, segregation of the Taylor vortex flow and Couette flow was observed, as shown in **Figure 9a**. The temperature profile (shown in **Figure 9b**) indicates that the temperature fluctuated in the Taylor vortex flow regime (T_{bl} and T_b) while that in the Couette flow (T_{bu}) showed no fluctuation. Because the viscosity in the lower part was lower than that in the upper part due to the axial temperature distribution, the local Re in the lower part was considered lower

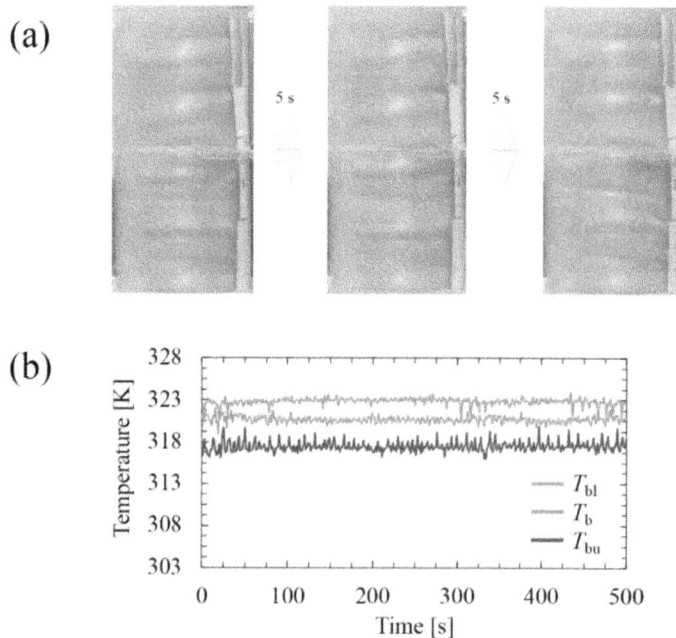

Figure 8.
Thermo-fluid dynamics of transient mode (Case IV) at C = 70 wt%, Re = 121.6, and Gr = 4653.5: (a) flow pattern and (b) temperature change with time [21].

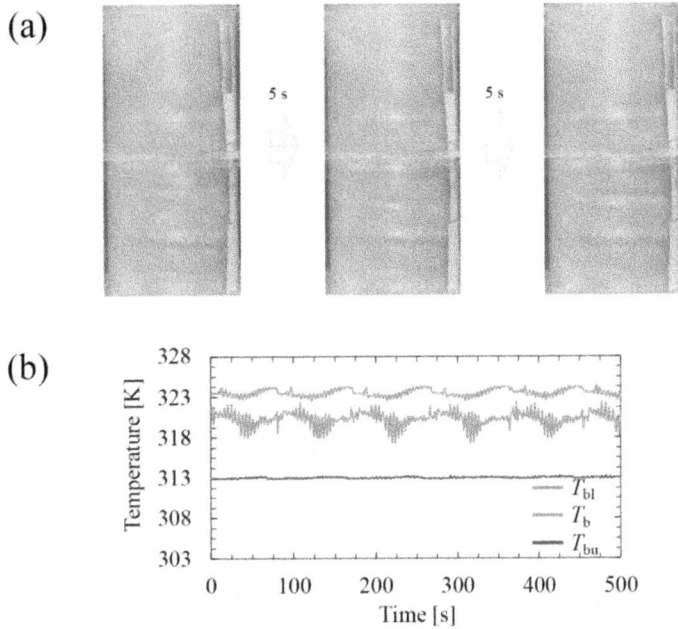

Figure 9.
Thermo-fluid dynamics of transient mode (Case V) at C = 80 wt%, Re = 47.5, and Gr = 492.2: (a) flow pattern and (b) temperature change with time [21].

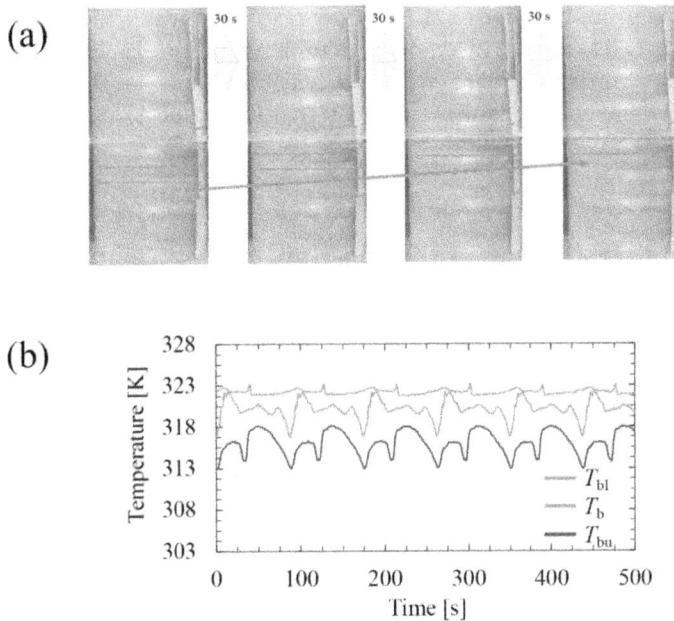

Figure 10.
Thermo-fluid dynamics of transient mode (Case VI) at C = 80 wt%, Re = 55.4, and Gr = 492.2: (a) flow pattern and (b) temperature change with time. The red square in (a) traces an upward of the inflow boundary [21].

than that in the upper part. It is expected that this distribution of local *Re* triggered the flow segregation. The flow pattern with flow segregation is referred to as Case V. With a further increase in *Re* at *C* = 80 wt%, an odd flow pattern was observed.

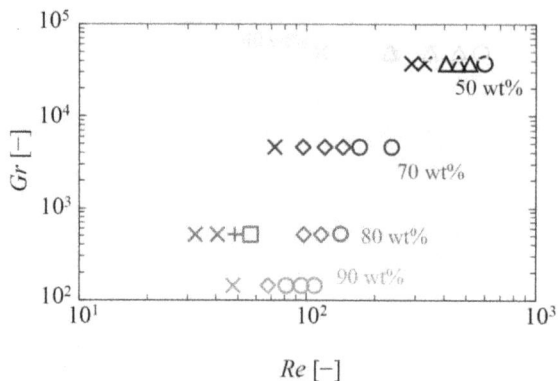

Figure 11.
Flow map based on Re and Gr. Black, blue, green, and red symbols show the case of C = 50, 70, 80, and 90 wt%, respectively, obtained in this study [21]. A part of data at C = 40 wt% is also plotted as a reference (denoted by yellow symbols) [20]. ×: Case I, △: Case II, ○: Case III, +: Case IV, □: Case V, ◇: Case VI.

Figure 10 shows the flow pattern and temperature change with time at Re = 56.7. As shown in **Figure 10a**, Taylor cells moved upward. The red square in **Figure 10a** traces the upward motion of a Taylor cell. As shown in **Figure 10b**, T_b and T_{bu} fluctuated due to the upward motion. This flow pattern with an upward motion of Taylor cells is denoted as Case VI.

The flow pattern for each C was classified based on Re and Gr. The map is shown in **Figure 11**. As indicated in **Figure 11**, the flow transition was affected by not only Re ang Gr but also C. This result infers that, to understand the Taylor-Couette flow dynamics in the non-isothermal system, other parameters except for Re and Gr are necessary. The detailed mechanism of flow transition is one of the interesting topics in future work.

5. Numerical investigation of thermo-fluid dynamics in non-isothermal system

As described in Section 4, it was inferred that heat/mass transfer is enhanced in Case II. Therefore, we investigated the fluid flow and heat transfer in Case II using numerical simulation. In particular, the local temperature/velocity was temporally detected at two positions: (r^*, θ^*, z^*) = P1 (0.5, 0, 0.26), and P2 (0.5, 0, 0.74). It is noted that r^*, θ^*, and z^* are non-dimensionalized coordinates of d, 2π, and $2h$, respectively. **Figure 12** shows the temperature change with time at Re = 516.4 and Gr = 21126.5. It was found that a similar pattern of temperature fluctuation is numerically reproduced in this condition. The temperature field and isosurfaces of axial velocity at Re = 516.4 and Gr = 21126.5 are shown in **Figure 13**. At t = 430 s, the cellular structure of Taylor cells was confirmed although it was deformed. As shown in **Figure 12**, the temperature difference at P1 and P2 was gradually developing at t = 430 s. Thereafter, at t = 480 s, heat convection mainly appeared instead of Taylor cells (**Figure 13**). As shown in **Figure 12**, the temperature at P1 and P2 quickly approached the approximately same value. Therefore, when switching from the Taylor vortex flow to heat convection, the rapid heat transfer occurs in Case II.

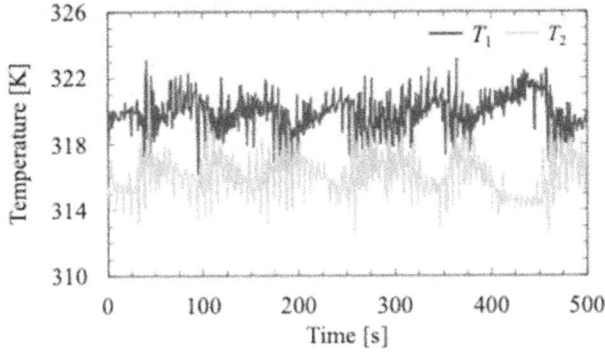

Figure 12.
Effect of degree of axial temperature gradient on temperature change with time at Re = 516.4 [21].

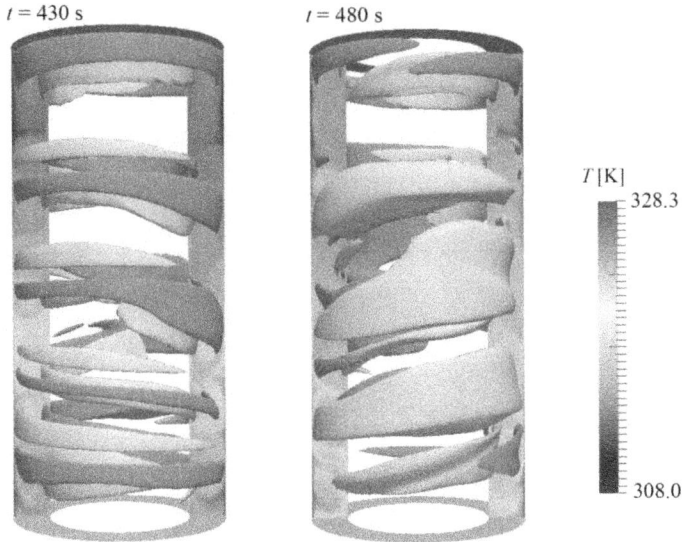

Figure 13.
Temperature distribution and isosurfaces of axial velocity (u_z) at Re = 516.4, a = 20 (Gr = 21126.5) [21].
Blue: isovalue u_z = −0.014 m/s, yellow: isovalue u_z = 0.014 m/s.

To quantitatively investigate heat transfer, an average Nusselt number at the outer cylinder surface, Nu_{av}, was obtained from the numerical results. Nu_{av} was defined, as follows:

$$Nu_{av} = \frac{1}{S}\int_S Nu_L dz \qquad (8)$$

where S is the heat transfer area and Nu_L is the local Nusselt number. Nu_L was calculated, as follows:

$$Nu_L = \frac{2hd}{\kappa} \qquad (9)$$

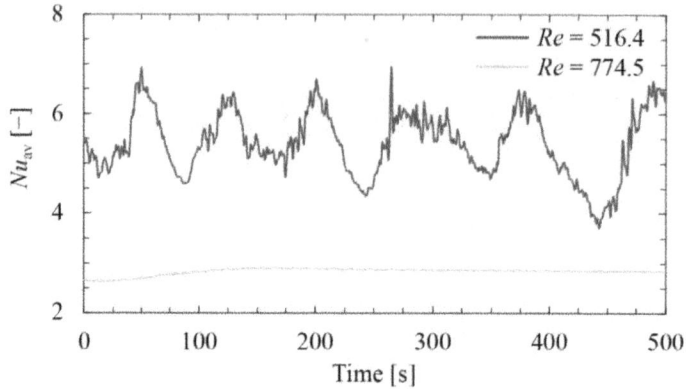

Figure 14.
Change in Nu_{av} with time at Gr = 27126.5.

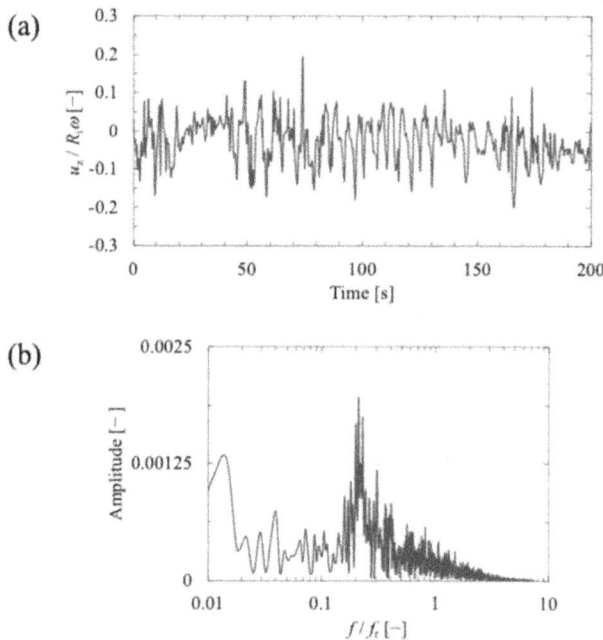

Figure 15.
Velocity characteristics at Re = 516.4 and Gr = 21126.5: (a) the fluctuation of axial velocity at P1 and (b) the analyzed result of fast Fourier transform (FFT) [21].

where h is the local heat transfer coefficient at the outer cylinder surface.
Figure 14 shows the Nu_{av} change with time at Re = 516.4 and 774.5. At Re = 774.5, the flow pattern corresponded to Case III (stable Taylor vortex flow). As indicated in **Figure 14**, the Nu_{av} at Re = 516.4 fluctuated with time due to the characteristics of the flow pattern in Case II. As described above, the Nu_{av} became the local maximum value when switching from the Taylor vortex flow and heat convection (e.g., t = 480 s). Remarkably, the Nu_{av} is enhanced at Re = 516.4 in Case II despite the lower Re than Re = 774.5 (Case III). This result suggests that the heat transfer augmentation is realized with the lower power input if the interaction of the Taylor vortex flow and heat convection is adequately utilized.

Finally, **Figure 15** shows (a) the fluctuation of axial velocity at P1 and (b) the analyzed result of the fast Fourier transform (FFT). As shown in **Figure 15a**, the velocity fluctuation is quite complicated. The FFT result (**Figure 15b**) indicates that the velocity fluctuation has a lot of frequency components. In the future, further investigation of velocity fluctuation would be useful for understanding Taylor-Couette flow in the non-isothermal system.

6. Conclusion

This chapter reviewed the Taylor-Couette flow dynamics in the non-isothermal system. We investigated the fluid flow and heat transfer experimentally and numerically. In experiments varying the concentration of glycerol aqueous solution, we observed six types of flow patterns (Case I–VI) and created the flow map based on Reynolds number (Re) and Grashof number (Gr). Several cases were shown in the specified concentration systems. This implies that the flow pattern cannot be classified only by Re and Gr. In Case II, a unique flow pattern was observed: the Taylor vortex flow and heat convection alternatively appeared. Based on the visualization of cross-sectional view and time series data of temperature in Case II, the enhancement of heat/mass transfer was expected when switching from the Taylor vortex flow and heat convection.

We also demonstrated numerical results for thermo-fluid dynamics in Case II. By numerical simulation, the flow pattern in Case II was successfully reproduced. Furthermore, the average Nusselt number in Case II was increased compared with the stable Taylor vortex regime (higher Re case). This result indicates that heat transfer augmentation is expected with lower power input in Case II. In addition, we showed the complicated velocity fluctuation in Case II. If the velocity fluctuation is investigated in more detail, we may extract a key topic to realize heat transfer augmentation under relatively lower Re conditions.

Acknowledgements

This study presented in this chapter was partially supported by JSPS KAKENHI (grant numbers JP18H03853, JP19KK0127, JP21K14450, 21KK0261, 23 K13592, and 24H00396).

Author details

Hayato Masuda[1]* and Naoto Ohmura[2]

1 Department of Mechanical Engineering, Graduate School of Engineering, Osaka Metropolitan University, Osaka, Japan

2 Department of Chemical Science and Engineering, Graduate School of Engineering, Kobe University, Hyogo, Japan

*Address all correspondence to: hayato-masuda@omu.ac.jp

IntechOpen

References

[1] Masuda H. Enhancement of heat transfer using Taylor vortices in thermal processing for food process intensification. In: A Glance at Food Processing Applications. Rijeka, Croatia: IntechOpen; 2021. DOI: 10.5772/intechopen.99443

[2] Coles D. Transition in circular Couette flow. Journal of Fluid Mechanics. 1965;**21**(3):385-425. DOI: 10.1017/S0022112065000241

[3] Kogelman S, DiPrima RC. Stability of spatially periodic supercritical flows in hydrodynamics. Physics of Fluids. 1970; **13**:1-11. DOI: 10.1063/1.1692775

[4] Grossmann S, Lohse D, Sun C. High–Reynolds number Taylor-Couette turbulence. Annual Review of Fluid Mechanics. 2016;**48**:53-80. DOI: 10.1146/annurev-fluid-122414-034353

[5] Kataoka K, Doi H, Kongo T, Futagawa M. Ideal plug-flow properties of Taylor vortex flow. Journal of Chemical Engineering of Japan. 1975; **8**(6):472-476. DOI: 10.1252/jcej.8.472

[6] Dusting J, Balabani S. Mixing in a Taylor–Couette reactor in the non-wavy flow regime. Chemical Engineering Science. 2009;**64**(13):3103-3111. DOI: 10.1016/j.ces.2009.03.046

[7] Kataoka K, Doi H, Komai T. Heat/mass transfer in Taylor vortex flow with constant axial flow rates. International Journal of Heat Mass Transfer. 1977; **20**(1):57-63. DOI: 10.1016/0017-9310 (77)90084-9

[8] Kataoka K, Ohmura N, Kouzu M, Simamura Y, Okubo M. Emulsion polymerization of styrene in a continuous Taylor vortex flow reactor. Chemical Engineering Science. 1995; **50**(9):1409-1413-1415-1416. DOI: 10.1016/0009-2509(94)00515-S

[9] Wang B, Tao S. Synthesis of micro-/nanohydroxyapatite assisted by the Taylor–Couette flow reactor. ACS Omega. 2022;7:44057-44064. DOI: 10.1021/acsomega.2c05491

[10] Masuda H, Horie T, Hubacz R, Ohmura N. Process intensification of continuous starch hydrolysis with a Couette–Taylor flow reactor. Chemical Engineering Research and Design. 2013; **91**(11):22592264. DOI: 10.1016/j.cherd. 2013.08.026

[11] Matsumoto M, Masuda H, Hubacz R, Horie T, Iyota H, Shimoyamada M, et al. Enzymatic starch hydrolysis performance of Taylor-Couette flow reactor with ribbed inner cylinder. Chemical Engineering Science. 2021;**231**:116270. DOI: 10.1016/j. ces.2020.116270

[12] Wereley ST, Lueptow RM. Inertial particle motion in a Taylor Couette rotating filter. Physics of Fluids. 1999;**11**(2):325-333. DOI: 10.1063/1.869882

[13] Ball KS, Farouk B, Dixit VC. An experimental study of heat transfer in a vertical annulus with a rotating inner cylinder. International Journal of Heat and Mass Transfer. 1989;**32**(8):1517-1527. DOI: 10.1016/0017-9310(89)90073-2

[14] Kuo D-C, Ball KS. Taylor–Couette flow with buoyancy: Onset of spiral flow. Physics of Fluids. 1997;**9**:2872-2884. DOI: 10.1063/1.869400

[15] White JM, Muller SJ. Viscous heating and the stability of Newtonian and viscoelastic Taylor-Couette flows.

Physical Review Letters. 2000;**84**:5130.
DOI: 10.1103/PhysRevLett.84.5130

[16] White JM, Muller SJ. Experimental
studies on the stability of Newtonian
Taylor–Couette flow in the presence of
viscous heating. Journal of Fluid
Mechanics. 2002;**462**:133-159.
DOI: 10.1017/S0022112002008443

[17] White JM, Muller SJ. Experimental
studies on the effect of viscous heating
on the hydrodynamic stability of
viscoelastic Taylor–Couette flow. Journal
of Rheology. 2003;**47**(6):1467-1492.
DOI: 10.1122/1.1621423

[18] Kaminoyama M, Nishimura K,
Nishi K, Kamiwano M. Numerical analysis
of bulk thermal styrene polymerization in
stirred vessel with double helical ribbon
impeller. Kagaku Kogaku Ronbushu. 1997;
23(6):835843. DOI: 10.1252/
kakoronbunshu.23.835

[19] Leng X-Y, Krasnov D, Li B-W,
Zhong J-Q. Flow structures and heat
transport in Taylor–Couette systems
with axial temperature gradient. Journal
of Fluid Mechanics. 2021;**920**:A42.
DOI: 10.1017/jfm.2021.430

[20] Masuda H, Yoshida S, Horie T,
Ohmura N, Shimoyamada M. Flow
dynamics in Taylor-Couette flow reactor
with axial distribution of temperature.
AIChE Journal. 2018;**64**(3):1075-1082.
DOI: 10.1002/aic.15972

[21] Masuda H, Nakagawa K, Iyota H,
Wang S, Ohmura N. Thermo-fluid
dynamics and synergistic enhancement of
heat transfer by interaction between
Taylor-Couette flow and heat convection.
Philosophical Transactions of the Royal
Society A-Mathematical Physical and
Engineering Sciences. 2023;**381**:20220116.
DOI: 10.1098/rsta.2022.0116

[22] JSME. Data Book: Heat Transfer. 5th
ed. Tokyo, Japan: Maruzen; 2009

www.ingramcontent.com/pod-product-compliance
Lightning Source LLC
Chambersburg PA
CBHW081336190326
41458CB00018B/6018